Learning to Teach and
Teaching to Learn Mathematics

Resources for
Professional Development

©2002 by the Mathematical Association of America (Inc.)

ISBN: 0-88385-168-7

Library of Congress Catalog Card Number 2001097389

Printed in the United States of America

Current Printing

10 9 8 7 6 5 4 3 2 1

Learning to Teach and Teaching to Learn Mathematics

Resources for
Professional Development

By

Matt DeLong
Taylor University

Dale Winter
Harvard University

Published and Distributed by
The Mathematical Association of America

The MAA Notes Series, started in 1982, addresses a broad range of topics and themes of interest to all who are involved with undergraduate mathematics. The volumes in this series are readable, informative, and useful, and help the mathematical community keep up with developments of importance to mathematics.

MAA Notes

MAA Service Center
P. O. Box 91112
Washington, DC 20090-1112
800-331-1622   fax: 301-206-9789

Preface

College and university mathematics departments shoulder a wide range of responsibilities and duties. Perhaps two of the most fundamental missions of any mathematics department are

1. the creation (or discovery) and preservation of mathematical knowledge, and

2. the communication of mathematical knowledge (to other professional mathematicians, students and the public).

The purpose of this book is to describe a set of tools and experiences for helping mathematicians to develop and enhance their instructional skills. By instructional skills we mean the ability to represent and communicate mathematical ideas and information to people who are not professional mathematicians. While creating this book, we have focussed on the challenges found in the teaching of undergraduate mathematics—especially the "introductory" classes taken by students at almost every college and university.

An important premise of our work is expressed by our belief that experienced mathematicians have important and unique contributions to make to the development of new mathematics instructors—contributions that cannot always be found in the (necessarily) more generic advice expressed through writings about college teaching and learning in general. As such, this book has been written for use by members of mathematics departments who are accomplished teachers but who do not necessarily have an extensive background in, or experience with, instructor development.

We have used various versions of these training experiences to address the developmental needs of graduate student instructors, post-doctoral and new regular-rank faculty, as well as adjunct and part-time faculty who were new to their institution. In our experience, with

- the provision of appropriate training experiences,

- support and advice from experienced instructors, and

- clear ideas concerning the nature and objectives of the course they are teaching,

instructors with little or no prior teaching experience can quickly become competent (and sometimes very skillful) teachers.

This book describes a semester-long program of professional development. It is intended to build upon an intensive pre-semester orientation course, such as the one described in [103]. The semester-long program integrates different forums for the examination, discussion and refinement of instructional practices. In our experience, this integration of multiple forums can enable training personnel to precisely tailor their advice, support and efforts to address the specific needs of individual instructors. For example, through the forum of a weekly training meeting, inexperienced instructors can be familiarized with fundamental "nuts and bolts" issues of running a classroom, while at the same time more experienced instructors can participate in discussions and evaluations of advanced philosophical and scholarly writings on learning and teaching through an educational issues seminar.

Over the last two decades, innovative instructional methods and powerful technological tools have become more commonly used in college mathematics courses, especially the "introductory" courses such as college algebra, precalculus, calculus, differential equations and linear algebra. The materials and training experiences described in this book are all highly compatible with these newer instructional methods. We have attempted to include a selection of materials that we believe will complement the excellent references on the teaching of university-level mathematics that are already well-known and widely available.

The contents of this book can be divided into five broad categories:

1. ideas for the structure of an integrated program of professional development (Chapter 1 and Appendix B),

2. suggestions on how to use the materials in the book (Chapter 2),

3. a description of (and supporting materials for) a brief pre-semester orientation session (Chapter 3),

4. support for a program of weekly training meetings (Chapters 4—16 and Appendix A), and

5. procedures and forms for conducting a system of class visits and feedback (Chapter 17).

Notes on 3—5 are given below.

Ideas on the structure of a professional development program

We believe that professional development is an ongoing process; one is not at the end of the professional development road after a one-semester training program. A successful, rich, and ongoing program of professional development will include many different forums for discussion and learning. To support this vision, we describe in Chapter 1 a fully-integrated model of such a system. We hope that this model will serve as a starting point for other departments who wish to do more in the way of professional development.

Support for a pre-semester orientation session

In Chapter 3 we describe a pre-semester orientation session. This session is designed to last for one day, and it is divided into three blocks.

Block A is intended for all new instructors and introduces the teaching resources, traditions and conventions of your department. Block A addresses issues such as the goals of the courses in your department, assessment of student work and course administration.

Block B is designed for beginning instructors and serves as an introduction to the basic ideas and methods of student-centered instruction.

Block C is primarily for instructors with little or no classroom teaching experience. This session provides an opportunity for new instructors to practice teaching, be videotaped and to receive feedback on their teaching.

The short sessions described in Chapter 3 may also be viewed as initiating discussions of teaching issues that will be resumed in the training meetings planned for the rest of the semester and described in Chapters 4 to 15.

Support for a program of weekly training meetings

The program of training meetings in Chapters 4–15 combines training in the use of nontraditional techniques of classroom activity, such as cooperative group work and active student participation in class, with the kinds of "nuts and bolts" issues that beginning instructors need to know, such as grading homework, proctoring exams, and working with students in office hours. The selection of subjects for the training meetings reflects our experience that many novice instructors can quickly become competent users of nontraditional teaching techniques. Simultaneously, they can acquire the knowledge of the more mundane responsibilities of instruction. It is assumed that participants in the training meetings have already finished preliminary training, such as the orientation session described in Chapter 3, or the course detailed in [103], including sessions on setting up a course, giving lectures, answering questions, and leading cooperative activities.

Each training meeting begins with an introduction to the content of the meeting that includes our rationale for advocating the teaching methods described and our reasons for structuring the meetings as we did. Each meeting is then described in detail, and many include extensive collections of background and supporting materials for distribution during the training sessions. We intend that meeting leaders will be able to run an effective training meeting with a minimum of preparation, by closely following the suggestions in the meeting chapters. The chapter for each training session is organized into the sections described below.

1. **Description and purpose of the meeting:** A very brief outline of what the meeting is about, and what participants will get out of it.

2. **Goals for the meeting:** A list of the training objectives for the meeting.

3. **Preparations for the meeting:** A checklist for the meeting leader to ensure that all preparations are carried out.

4. **Agenda for the meeting (to be distributed to participants):** A list of the topics that will be dealt with during the course of the meeting.

5. **Outline for running the meeting:** A detailed description of the meeting, including a "time line" for meeting leaders to follow, and full instructions on what to use each segment of meeting time for. (Some meetings include several different outlines, each reflecting a different format for the meeting, so that meeting leaders have more flexibility in how they will conduct the meeting.)

6. **Ideas for expediting the meeting:** Suggestions for cutting corners in the interest of saving meeting time.

7. **Suggested reading:** A short annotated bibliography that points the meeting leader to other sources relevant to the topic treated in the meeting. These readings can be used by the meeting leader in preparation for the meeting, or they can be suggested to the meeting participants as follow up.

8. **Meeting materials:** Handouts and background material for distribution during the meeting.

Since the range of experiences that we describe in this book will not be completely consistent with the goals and methods of every professional development program, there will be teaching issues that we have not substantially addressed, but which are nonetheless highly relevant to the reader. To address this need, in Chapter 16 we provide a framework for adapting materials to the specific needs of other programs. There we describe the development of an additional training meeting, but our emphasis is less on the content of the particular meeting and more on the method that we use for developing new meetings from scratch.

Finally, Appendix A is devoted to planning and running successful meetings. This appendix identifies common criticisms of meetings, and suggests ways that these difficulties can be addressed. The appendix describes methods for running meetings—such as assigning roles—and indicates how these methods can be successfully implemented in the academic setting. It also describes examples of difficult or disruptive participant behavior, and suggests numerous strategies for dealing with these "people problems." Finally, this appendix gives a method for meeting improvement, together with a questionnaire, and suggests an alternative to training meetings that can be used to continue the professional development of instructors who have "graduated" from the program of training meetings.

Support for a system of class visits and feedback

In Chapter 17 we describe different kinds of classroom visits–observational visits, student feedback sessions, and peer visits–that we have used while training instructors—ranging from beginning graduate students to senior faculty—at the University of Michigan, Duke University, Taylor University and Harvard University. We describe (in considerable detail) procedures for conducting each kind of visit, including how to set up visits, what to do while in the classroom, how to write a report, and how to conduct the subsequent meeting with the instructor visited. In addition, we list the advantages and disadvantages of each kind of visit. We include a plan for a semester-long program of class visits, along with a large amount of practical wisdom concerning instructors' reactions to visits, and many of the problems that can occur.

Acknowledgments

The contents of this book represent seven years of collective experience in professional development at the University of Michigan, Duke University, Taylor University and Harvard University. As can be well expected, many individuals have contributed in various ways to this program. We wish to acknowledge the program heritage and the wealth of help and guidance provided by Beverly Black, Morton Brown, Robert Megginson, and Patricia Shure. Some of the meeting materials in "Homework Teams" were produced in collaboration with William Cherry, Doug McCulloch and Marius Irgens. Some of the meeting materials in "Administrative Issues for the End of Semester" were produced in collaboration with Glen Whitney and Donatella Delfino. The writing workshop in "Adapting Materials and Designing Your Own Meetings" was developed with William Hoese. We also wish to acknowledge the staff of the Center for Research on Learning and Teaching at the University of Michigan, and in particular, Beverly Black and Matt Kaplan who provided training in classroom observation and the collection of student feedback, and Barbara Hofer who conducted focus groups with mathematics instructors in 1995. We acknowledge the instructors whom we have had the pleasure of training, and from whom we have learned much about teaching mathematics as well as training others to teach mathematics. We thank Martha Gach and Tim Hsu for reading and commenting on the first draft of this book. We also thank Chris Swanson and Anton Kim who used an earlier version of these materials at the University of Michigan and provided comments on some of the chapters. Finally, we would like to thank Tina Straley, Barbara Reynolds and Paul Fishback and the other members of the MAA Notes editorial board for their support of this project and for their comments on earlier versions of our manuscript.

The preparation of this book was partially supported by a 1997 Rackham Pedagogy Grant and a 1998 Rackham Pedagogy Grant.

Contents

Chapter 1

The Professional Development Program

1.1 Overview

We believe that it is important for a mathematics department to be actively involved in the professional development of its faculty and graduate students. By *professional development* we mean the training, assessment, and improvement of teaching skills and practices. Professional development that comes from within the mathematics department, rather than through non-disciplinary channels, is more likely to be meaningful and credible to the instructors who are involved. For example, sensibilities about teaching mathematics are sometimes quite different from those in other disciplines. Compare what is meant by "read the textbook" in a mathematics class to what is meant by the same statement in a literature or history class. In addition, the types and standards of argumentation and precision are different in the mathematics department than they are elsewhere. Moreover, there are certain obstacles that are more prevalent in mathematics than elsewhere. For instance, innumeracy and mathematical illiteracy are far more socially acceptable, or even favored, than is an inability to read or write. For these and other reasons, we feel that the best place for training and supporting mathematics instructors is within the mathematics department.

We believe that an effective way for a mathematics department to train its inexperienced instructors, and to support and develop its other faculty members and graduate students is through an integrated professional development program that includes four components. By "inexperienced instructors" we include not only graduate student teaching assistants who are new to teaching, but also post-doctoral and other new faculty members who are gaining experience as instructors, and also adjunct or part-time faculty who have more teaching experience, but may be new to teaching in your department. The four components we envision are a pre-semester training course; weekly professional development meetings; class visits and feedback from supervisors, students, and peers; and a departmental educational issues seminar. Such a program would

- adequately equip new instructors for their first experiences teaching collegiate mathematics,

- continue to support instructors as they grow in their teaching philosophies and abilities,

- increase the value of teaching in the department,

- help to create a community of teachers, and

- help to unify a departmental teaching vision, if one is not already established.

Such a program could be assessed using teaching portfolios, which, in addition, would be useful to graduate students and non-tenure track faculty as they approach the job market and regular-rank faculty as they approach tenure and promotion decisions.

This chapter contains a brief overview of the four components of an integrated professional development program, an outline of how teaching portfolios can be used in conjunction with such a program, and ideas on continuing professional development. Each section points to more detailed information on implementing that aspect of the program. Figure 1.1 is a schematic of how the four components relate, and how they are bound together via teaching portfolios. The relationships enumerated in the figure are explained below.

1. Feedback from class visits alerts instructors to issues in pedagogy or practice about which they may like to find out more.

2. Feedback from class visits indicates what instructors are struggling with, and what course coordinators should focus on in meetings.

3. Weekly meetings provide a forum for treating teaching topics that are not relevant to the first few weeks of the semester.

4. Pre-semester training provides a foundation for further training during weekly meetings.

5. The educational issues seminar provides a forum for discussing teaching topics that were not addressed in training, and it provides coordinators with further material for future versions of training.

6. Pre-semester training identifies areas of instructor interest, thereby suggesting topics for the educational issues seminar.

7. The educational issues seminar provides ideas for new topics to be discussed during the weekly meetings.

8. Class visits confirm that the instructors are doing their jobs properly, and they supply ideas for use during future versions of training.

9. Weekly meetings provide the skills and knowledge that instructors need in order to participate in a seminar on educational issues.

10. Portfolios provide motivation for instructors to do excellent work.

11. Participation in a wide range of professional development activities guarantees that there will be artifacts to include in a teaching portfolio.

1.2   Pre-Semester Training

The first component of an integrated professional development program is a pre-semester training course primarily intended for those new to teaching or those new to teaching in the style of the department, although more experienced instructors may benefit from participating or assisting in the planning and implementation of the course. The course is likely to be short but intensive, covering the most fundamental issues in the limited amount of time. The goals of the pre-semester training course are to prepare the new instructors to begin the semester, to equip them for handling the immediate teaching experiences of the first few weeks, and to begin a foundation of teaching philosophies and skills on which to build through further professional development activities.

The pre-semester training may include:

- an introduction to the departmental teaching philosophy;

- information on departmental and institutional administrative policies;

- examples of the handouts needed to start the semester;

- training on setting the groundwork for active learning and cooperative learning, getting students to read the textbook, and using graphing calculators in the course;

- facilitated sessions concentrating on practice lecturing, answering questions, and running cooperative activities in class;

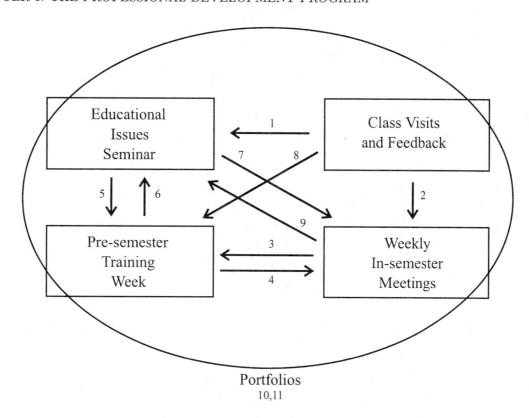

Figure 1.1: The Components of an Integrated Professional Development Program

- support for planning and managing daily classroom activities; and

- information on departmental and institutional policies on professional responsibilities and ethics.

Some institutions have implemented a one-week or longer pre-semester training course. The details of one such course have been documented by Shure, Black and Shaw [103]. Although a week-long program can cover more ground (in more detail) than a short orientation session, some institutions may find themselves unable to commit sufficient resources and time to organizing and running an elaborate pre-semester training course. For such institutions, Chapter 3 of this book details a much shorter training session.

The pre-semester training course is integrated into the professional development program by giving course organizers a sense of what would be most beneficial for the specific group of new instructors to work on during the in-semester training sessions, by giving course organizers and other instructional consultants a sense of what to look for during class visits, and by showing departmental personnel possible topics of interest for discussion in the educational issues seminar.

1.3   Weekly Development Meetings

The second component of an integrated professional development program builds upon the pre-semester training course through weekly development meetings. These meetings are again intended primarily for the less experienced instructors of the department, although other instructors can choose to attend those meetings of particular interest to them. The experienced instructors can also find substantial benefit from assisting during the meetings. The training meetings provide a natural forum for experienced instructors to share their experiences with new instructors. The goals of the weekly development meetings are to provide more detailed training on some of the topics encountered in the pre-semester training course, to provide training on topics that are difficult to address without the participants first having some experience in the classroom, and to dispense information and training on situations that are not encountered until later in a semester.

The in-semester training meetings may cover:

- topics encountered in the pre-semester training that require more detailed coverage:

  - using groups in class,

  - getting students to read the textbook,

  - teaching with technology, and

  - writing lesson plans;

- topics that are better addressed after instructors have some classroom experience to frame the discussion:

  - control in the classroom,

  - difficult students,

  - student motivation, and

  - office hours;

- topics that are not usually encountered until later in a semester:

  - grading and writing quizzes,

  - proctoring exams,

  - changing homework teams, and

  - end of semester issues.

The bulk of this handbook is devoted to the implementation of a program of in-semester training meetings that build upon a pre-semester training course like the one detailed in [103].

1.4   Class Visits and Feedback

An additional component of an integrated professional development program is a system of classroom visits and feedback. These visits are most crucial for inexperienced instructors, but instructors at any stage of their careers can benefit from the observation and feedback process. The class visits can take the form of observational visits by course directors or other instructional consultants, facilitated student feedback, or peer visits. Through class visits, instructors can get concrete, focused, and personal feedback. This feedback can build upon and solidify the training and ideas encountered during the development sessions.

A system of classroom visits can benefit the organizers of the professional development program as well as the instructors being visited. Through the sharing of sensible and helpful advice, the course coordinators can gain the respect and trust of their instructors. In addition, visits keep coordinators connected with the classrooms, and the needs of the instructors whom they are training.

The different types of class visits are related to each other. Observational visits can inform facilitators of what to emphasize during student feedback visits and peer visits. They can also identify good candidates for participation in a peer visit program. Participation in a program of peer visits can provide valuable experience to those instructors who are interested in assisting with future versions of the professional development program.

Chapter 17 of this book contains information on implementing a program of classroom visits.

1.5   Educational Issues Seminar

The final component of an integrated professional development program is a departmental educational issues seminar. New instructors who are heavily involved in the other three aspects of the professional development program are not likely to have the time for regular participation in such a seminar. However, a seminar can be used to maintain the momentum new instructors have received through training and class visits, and to encourage dialogue among instructors of vastly different experience levels.

In addition to building upon the foundation of professional development, some other goals for an educational issues seminar in a department of mathematics may include:

- fostering a collegial community approach to teaching,

- raising the value of teaching in the department,

- forging a unified vision for the department's teaching philosophies and practices,

- fostering conversations on pedagogy and practice that will continue informally,

- refreshing the teaching skills and methods of more experienced instructors, and

- creating more meaningful connections between research and classroom teaching.

For example, in 1997–1998 The University of Michigan Department of Mathematics instituted their own Educational Issues Seminar, with many of these goals in mind. The seminar met biweekly. The likely participants were polled as to their interest in particular topics. The organizers then chose topics, provided short related readings in advance, and led a semiformal discussion during the one-hour meeting. To encourage attendance, lunch was served. Attendees ranged from new Graduate Student Instructors to tenured faculty in top administrative positions.

More ideas on the implementation of a departmental seminar on educational issues in mathematics can be found in [19].

1.6   Using Portfolios

Teaching portfolios can be used to unify the professional development program, by documenting the experiences and growth of the instructors involved. A portfolio captures as concrete artifacts what an instructor does when he or she is teaching, in such forms as instructional materials, reflective essays, peer and supervisor reviews, student ratings and testimonies, videos, and student work. These artifacts can be used for personal reflection, community sharing, and departmental evaluation. The portfolio documents an instructor's participation in the different components of the professional development program and demonstrates how his or her teaching philosophy and practice have been affected. Building a teaching portfolio helps an instructor to maintain a spirit of professional development by encouraging continued reflection on teaching decisions and outcomes.

The items that may be included in a teaching portfolio are many and varied. The following list is a condensed version of that found in [116]. Many of these items will be naturally produced through the four components of an integrated professional development program listed above. The portfolio can document

- what we teach: content, goals;

- how we teach: skills, methods, innovations;

- change in our courses: improvements, variations;

- rigor in academic standards: evaluation philosophies, examples of evaluations, assessment of evaluations;

- student impressions: special feedback, ways you've responded to feedback, patterns of response;

- efforts at development: ways you've studied your teaching, workshops and seminars attended, sharing of ideas with others;

- assessment by colleagues: supervisor and peer reviews, how feedback is used to make changes, teaching awards and commendations.

The development of a teaching portfolio can be a basis for professional growth, even if it is not formally attached to a program of professional development within a department. In addition, it can serve as a valuable way of presenting an instructor's abilities for review by a potential or current employer.

1.7   Ongoing Professional Development

At some point, instructors will "graduate" from the formal parts of any professional development program. A successful professional development program will generate enthusiasm among instructors, many of whom will wish to participate in ongoing development.

Instructors who have "graduated" from pre-semester and in-semester training programs can continue development activities through participation in a departmental educational issues seminar, attendance at on-campus instructional workshops, attendance at workshops sponsored by professional societies such as the MAA or the AMS, implementation of self-development using teaching portfolios, and participation in focus group meetings.

Focus group meetings are explained in Appendix A. In an ongoing program of professional development, focus group meetings can provide a vehicle for experienced instructors to meet with their peers, and maintain a process of ongoing professional improvement. Focus group meetings can also be used in conjunction with a program of peer visits, which are detailed in Chapter 17

Chapter 2

How to Use this Book

This book is intended to be a handbook for implementing three of the components of an integrated professional development program described in the previous chapter. Chapter 3 details a short pre-semester training course, although those departments that are able to commit to a longer pre-semester training program may use the model provided by Shure, Black, and Shaw [103]. Chapters 4 through 16 are devoted to the weekly in-semester training meetings. Chapter 17 is devoted to class visits and feedback. This chapter explains how to use the rest of the handbook to implement these three components. More detailed advice on running effective meetings is provided in Appendix A. Detailed discussions of implementing an educational issues seminar and using portfolios to integrate a professional development program are not included in this book.

2.1 The Orientation Session

Chapter 3 describes a session suitable for introducing new instructors to their mathematics department, and for getting all instructors ready to teach the first week of class. Much of the format of that chapter is similar to the format of Chapters 4 through 15, which detail the twelve in-semester training meetings. Therefore, most of the instructions for using Chapter 3 can be gleaned from those given below.

2.2 The Meeting Chapters

This book contains chapters detailing twelve training meetings. Each chapter is divided into the following sections.

Description and Purpose of the Meeting: This is a brief description of the purpose of the meeting, and what is likely to take place if the meeting is run as recommended. There is also an estimate for the length of the training session. Usually this ranges from thirty to fifty minutes, and most of the meetings are intended to be fifty minute sessions.

Goals for the Meeting: This is a list of objectives that are likely to be realized if the meeting is run as recommended.

Preparation for the Meeting: This is a preparation checklist for the leader(s).

Agenda for the Meeting: This is an agenda of how the meeting will probably proceed. This agenda is suitable for distribution to participants in advance of the meeting.

Outline for Running the Meeting: In most cases, this provides a detailed description of exactly what the leader will need to do at each point of the meeting to achieve the goals of the meeting, and to complete the items listed on the agenda. Suggestions for the amount of time that should be spent on each part of the meeting are given here.

Meeting Materials: This is a checklist for the leader(s), and it lists the handouts and supporting materials that are provided at the end of the chapter or that need to be assembled before the meeting.

Suggested Reading: This is a short annotated list of suggested readings on the topic under consideration.

A Suggested Schedule for Meetings.

Here we provide a suggested order for scheduling the meetings in this handbook. In ordering these meetings we assumed a weekly meeting schedule during a semester of traditional length.

1. Making In-class Groups Work.

2. Getting Students to Read the Textbook.

3. Assessing and Evaluating Students' Work.

4. Managing Homework Teams.

5. Teaching During Office Hours.

6. Establishing and Maintaining Control in Your Classroom.

7. Proctoring Tests and Examinations.

8. Improving Teaching with Graphing Calculators and Computer Algebra Systems.

9. Making Lesson Plans.

10. Strategies for Motivating Students.

11. Dealing with Difficult Instructor-Student Situations.

12. End-of-Semester Administration.

The meetings have been ordered so as to deal with the most fundamental and most immediate skills first, while leaving for later issues that are more advanced or more likely to arise later in a semester. This may seem contrary to delaying the meeting "Making Lesson Plans" until later in the semester. We suggest that, in order to allow the new instructors to initially focus on their classroom practice, the course coordinators provide these instructors with fairly detailed sample lesson plans[1] until this meeting is held. The new instructors can then be encouraged to follow these lesson plans closely. After the meeting "Making Lesson Plans" is held, the instructors will be able to write their own lesson plans.

Although writing lesson plans for new instructors requires additional institutional support, it has the advantages that it provides an additional vehicle for reinforcing suggested student-centered teaching methodologies, and it allows instructors to develop a conception of their own classroom based on experience before they are required to write their own lessons. If your institution is not able to provide sample lesson plans to new instructors, then we suggest that you move the training meeting "Making Lesson Plans" to the first meeting of the semester.

Because teaching is a complex activity, there are too many topics that need to be addressed early in the semester of a training program for novice instructors. There are several ways to alleviate this crunch and you will need to find the one that fits your institution best. The approach that we advocate is to build initial skills sufficient for a few weeks of teaching by holding a pre-semester training program. Later in-semester meetings can then be used to refine these skills. Another approach that we also advocate is to provide additional support from the training staff to assist the novices with those issues that must be delayed to later meetings, but which they otherwise would have to address in their teaching. This is the approach that we suggested above for dealing with lesson plans. A third approach is to meet more often than weekly during the early portion of the semester, and then meet less often as the semester progresses. This may be appreciated by the novice instructors, as the time pressures only get worse as the semester wears on.

---

[1]Three samples of lesson plans that were provided to novice precalculus instructors at the University of Michigan are given in Appendix B.

In any case the actual order in which the meetings are held should reflect the needs of the instructors in the training program. For example, if many instructors lose control of their classes at the beginning of the semester, then it would make sense to hold the meeting, "Establishing and Maintaining Control in Your Classroom" as early in the semester as possible. Other meetings lend themselves to a particular time. For example, the meeting, "Proctoring Tests and Examinations" will be most effective if held immediately before the first major test or exam that the instructors will have to administer.

The Format of the Meetings

Meeting styles can be very formal, very informal, or somewhere in between. Organizers must decide at what level of formality to conduct a meeting based on their intentions for what will be accomplished during the meeting, and their perception of what the participants need to get out of the meeting.

The meeting formats presented in the chapters of this book represent meetings that have been used effectively. In most cases, the format presented is a reasonably formal training session, although the nature of some of the meetings required a more fluid and less formal approach.

You may find it necessary to adopt a format for some meetings that differs from the format suggested in this book. For the purposes of comparison, three different formats have been given for the meeting, "Dealing with Difficult Instructor-Student Situations." Readers are invited to compare how the goals of this meeting can be realized through different formats, and how the agenda can be tailored accordingly.

Using the Chapter to Prepare for a Meeting

The amount of information supplied in each chapter is substantial. Although it is sometimes possible for an exceptional and experienced leader to run an effective meeting with little or no preparation, this is not often the case.

Careful and thorough preparation is the best guarantee of an effective training meeting.

In particular, if you arrive at the meeting with no idea of the goals for the meeting or what needs to be done in order to achieve those goals, then the meeting will almost certainly fail.

Each meeting description contains a section entitled "Preparation for the Meeting." This section is included as a checklist for the leaders to ensure that all essential materials will be assembled for the meeting, and that participants will have been informed of the meeting. Simply carrying out the listed meeting preparations, without carefully reading the meeting chapter, is unlikely to result in an effective meeting. A method of preparation that is more likely to result in a successful meeting includes carefully reading the meeting chapter, gathering the meeting materials well in advance, and following the time-line for preparation given below.

Reading the Meeting Descriptions

It is imperative for you, and any other meeting "executives," to read the meeting introduction and description thoroughly *well before* the meeting. It is just as important that everyone at the meeting be well aware of the *purpose* and *goals* for the meeting. Therefore, you should first read the introduction, the description, and the goals for the meeting. If you find that some goals that are important to your training program are not listed in the chapter, then they can be added. The other parts of the meeting can be subsequently adjusted so that the meeting will also address these goals. You may also look at some of the suggested readings provided in the chapter in order to solidify your own perspective on the topic to be discussed.

The next section for you to read is the outline for running the meeting. The reason for reading this section out of sequence is to confirm that the format suggested for the meeting is one that you think will have an excellent chance of success. If this is not the case, then you will need to use the content of the suggested format to devise your own plan for running the meeting, and to adjust the preparations, agenda, and meeting materials accordingly.

Finally, the sections describing the preparations for the meeting and the agenda should be read and modified, if necessary. It is now assumed that you have decided on an appropriate format for the meeting, you have developed a plan for running the meeting, and you have formulated an agenda to be circulated to participants.

Meeting Materials

A checklist of the meeting materials is provided in each chapter. The chapters include handouts and supporting documents that you can choose to distribute before, during, or after the meeting. The meeting descriptions given in the book usually have specific uses for each of the different handouts. If the meeting descriptions given in this book are followed closely, then the supporting documents will be most useful if given either in advance of, or during the meeting.

You will undoubtedly find that some of the meeting materials included are not completely compatible with your institution's guidelines or customs. In such a case, you can produce a similar document to the one given that better reflects the institutional practices, or you can opt to distribute nothing at all. If you decide to distribute nothing, then the agenda and the outline for running the meeting will probably have to be modified to take this change into account.

A Time-Line for Preparing Meetings

The following time-line retains a great deal of flexibility, and you should adapt it to suit your situation.

Ten to fourteen days before the projected meeting: The meeting leader should meet with other "executives" to develop objectives for the training program for the next two weeks.

Ten to fourteen days before the projected meeting: The meeting leader should decide whether a meeting on a particular topic is actually necessary.

Seven to Ten days before the projected meeting: If a meeting is justified, the meeting leader should plan an agenda based on the suggestions from the meeting chapter and any extra goals the leader might have identified. The leader should decide who should attend the meeting.

One week before the projected meeting: The leader should find a room in which to hold the meeting. The leader should announce the meeting, including the date, time and venue for the meeting, and circulate the agenda to participants.

Five to Seven days before the projected meeting: The meeting leader should read the meeting description and make whatever modifications seem appropriate.

One to Five days before the projected meeting: The meeting leader should prepare the meeting materials.

The day of the meeting: The leader should reread the goals and outline for running the meeting, and then conduct the meeting.

One to three days after the meeting: If desired, the leader should follow up on the effectiveness of the meeting.

With the demands that most instructors feel on their time, participants will appreciate as much advance warning of a meeting as possible. However, if a meeting is announced too far in advance, some participants may forget about it. Uncooperative participants have been known to deliberately schedule other commitments to conflict with a meeting, even when well aware of the meeting time. Unfortunately there is not much that a meeting leader can do to deal with this situation, except perhaps trying to find another venue for relaying the truly vital information discussed in the meeting. All instructors appreciate advance warning and most consider one week to be the minimum amount of notice.

Starting and Ending Meetings

The meeting outlines usually call for some kind of introduction to each meeting. After thanking the participants for their attendance, an effective way to kick off the meeting is to reiterate the goals or objectives for the meeting; however, simply reciting a list of goals has little impact on many meeting participants. Writing the goals on a chalkboard or preparing an overhead transparency will be more effective.

During the introduction to the meeting is also a good time to hand out any institutional guidelines that may apply to the topic of the meeting. For example, during the meeting "Teaching During Office Hours," you may pass out a handout that specifies, among other things, how many office hours per week an instructor must hold, acceptable place to hold the "office" hours, how many *different* days and times an instructor must hold office hours, and the consequences for routinely shirking office hour duties.

Most of the meeting outlines suggest ending the meeting by summarizing important points, providing time for participants' questions, handing out suggestions for additional readings, or distributing questionnaires. Which of these you include will depend upon the type of meeting, and your goals for the meeting. Unless you are running ahead of the meeting outline, there will not be time in the five minutes suggested for all of these activities. If time runs short, you could ask the participants to bring their questions to you individually after the meeting, or you could pass out any summaries or other handouts as the instructors leave the room. In addition to these suggested activities, you should always thank the participants for their attention before they leave the meeting.

A questionnaire is provided in Appendix A. In practice, a questionnaire will probably only be necessary for the first few meetings. After two or three meetings, it is usually possible for you to develop an accurate picture of the likes and dislikes of the participants. As soon as the questionnaires are no longer developing new and useful information, it is probably time to stop using them.

Ideas for Expediting the Meetings

If you feel the need to run some of the meetings in an abbreviated fashion, there are some techniques that can serve to expedite a meeting. These include the following.

- Distribute any handouts to be read prior to the meeting.

- Some activities that are suggested in the meeting may be assigned to be done prior to the meeting, with only the discussion held during the meeting. For example, in the chapter "Making In-class Groups Work," the participants could be given the potential cooperative activities in a handout prior to the meeting, and the participants could be asked to analyze these and bring their ideas to the meeting.

- Some suggested discussions may be distilled by the meeting leaders into a handout to be given to the participants. However, it is likely that the participants will feel more ownership over ideas that they themselves bring to the meeting.

- If a meeting suggests having a panel of experienced instructors speak to the participants, this could be replaced by asking the experienced instructors to prepare a one-page handout describing their experiences.

- If a meeting suggests a role-play activity, have a team of experienced instructors role-play the activity for the entire group, or type up a vignette (such as the one given in the chapter "Teaching During Office Hours") for the participants to read and analyze.

- Questionnaires could be distributed to participants after the meeting, with the instructions that they be completed and returned on the participants' own time. If you use this strategy, be aware that this can significantly reduce the return rate of completed questionnaires.

- If the meeting's primary function is the dissemination of information (such as the chapter "Proctoring Tests and Examinations"), then do not hold a meeting. Instead, distribute institutional guidelines and handouts to the instructors, and invite them to bring to you any questions that they may have. If you use this strategy, be aware that not everyone takes the time to read materials that are simply handed to them! You may have some difficulties with instructors who do not read important information.

Questions for Meeting Leaders

- Deciding to have a meeting:

  - What are the goals or objectives for your training program for the next two weeks or so?

  – Is a large meeting really essential to achieving these goals?

  – Are there any meetings described in this book that can help you achieve these goals?

- Preparing for a meeting:

  – Have you done the reading for the meeting?

  – What meeting format will work well?

  – What should be on the agenda for the meeting?

  – Who needs to come to this meeting?

  – Do you have a time and day that you know most people will be able to attend?

  – Do you have a room for the meeting that is big enough and has the proper equipment?

  – Have you given participants enough warning of the meeting—a week at least?

  – What will you do about people who cannot attend the meeting?

  – Are the meeting materials from this book suitable? If not, how do you modify them?

  – Have you made enough copies of the materials for everyone?

  – Have you done everything in the section "Preparation for the Meeting?"

  – Have you reread the meeting description?

  – Do you have everything that you need for the meeting?

- Following up on a meeting:

  – What made the meeting effective?

  – What could make the next meeting more effective?

  – Have you followed up on questions that you were unable to answer during the meeting?

  – Have the people who missed the meeting been dealt with appropriately?

Developing New Meetings

If there are goals of your particular professional development program that are not addressed by the training meetings provided in this book, then Chapter 16 can be used to develop meetings that address these additional goals. To use this chapter, a training coordinator must first identify the additional goals to address. The training coordinator should then gather resource materials that address these goals. These materials could be those developed within the department, or they could be culled from the MAA notes series, journal articles, or the literature on good general classroom practices. Once the resource materials have been acquired, the trainer can read Chapter 16 and use the methods specified there to fashion the resources into a training meeting.

2.3   Classroom Visits

Chapter 17 explains three main types of class visits, and how they can be used to give feedback to individual instructors on their teaching, to personalize the professional development program, and to give feedback to the course coordinators on the effectiveness of the professional development.

   To use this chapter, a course coordinator should first read the section "Class Visits and Professional Development" in order to become familiar with the purposes of class visits, and how they can be integrated into a professional development program. The possible goals for class visits are listed in this section. The coordinator should read these goals, and modify them to meet his or her department's needs.

   Next the coordinator should read the following four sections of Chapter 17, which detail different types of visits. In particular, the coordinator should pay attention to the advantages and disadvantages of each type of visit. The coordinator should then weigh the goals for his or her department with the advantages and disadvantages of the types of visits in order to decide if class visits would be a useful component of the professional development program, and if so, to what extent they will be used. Based on this, the coordinator should develop a schedule for using the visits. The section "Ideas for Using Visits Throughout the Semester" provides an extensive framework for combining the visits, and it can be used as a starting point from which to develop a schedule of visits.

It is most desirable if all the instructors scheduled for a particular type of visit are observed in close succession. It is unreasonable for one facilitator to attend and give feedback for more than five classes per week. Therefore, if an extensive program of visits is planned, the course coordinator should assemble a team of visitors to assist with the observation and feedback.

Before implementing a program of visits, the course coordinator and the other facilitators should read Section 17.6 "Potential Problems with Implementing Class Visits." This section provides scenarios of potential difficulties, for which the team of visitors should mentally prepare responses.

Finally, as the time for a particular type of visit approaches, the coordinator and other visitors should divide up the instructors to be contacted. The team of facilitators should each re-read the particular section of Chapter 17 that is devoted to that type of visit. A more labor-intensive, but thorough, approach would be for the course coordinators to develop a workshop based on the material in this book for the facilitators, including practice observing and giving feedback to other facilitators (or observing and discussing videotapes of actual classes). The team of visitors should contact the instructors and implement the class visits. After the visits are concluded, the team of facilitators should report back to the course coordinator, maintaining confidentiality where appropriate, in order to provide information useful for customizing the current iteration, and streamlining future iterations, of the training program.

Chapter 3

An Orientation Session for the Beginning of the Semester

3.1   Introduction

The purpose of this chapter is to describe a session suitable for introducing new instructors to your mathematics program. As mentioned in Chapter 1 of this book, we envision a training program that includes intensive pre-semester training as an integral and important component. Throughout this book we have made reference to the pre-semester training materials developed by Shure, Black and Shaw [103] at the University of Michigan. Our experience suggests that such a pre-semester training program provides new instructors with a set of foundational experiences that the meetings described in Chapters 4–15 build upon.

Unfortunately, circumstances such as budget, availability of training personnel, and time sometimes make an intensive pre-semester training program a luxury that departments feel that they cannot afford. Recognizing this reality, we include this chapter that describes a very basic form of pre-semester training. The orientation sessions that we have described here raise many of the issues that will be expanded upon by in-semester training sessions, such as grading student work, using cooperative learning and teaching with technology. However, the sessions that we describe in this chapter should not be seen as a "stand alone" training program. They are intended to help instructors to recognize the issues that they will likely have to cope with on the first day or during the first week of class. We feel that it is very unlikely that the orientation session that we describe here—if offered in isolation, without a follow-up program of in-semester meetings, visits, and activities—will help instructors to become competent or masterful users of the teaching techniques and ideas that we describe.

There is considerable evidence to suggest that teachers tend to imitate their own teachers. With this in mind, it is important to try to make the orientation session as interactive as possible, and to actively involve the trainees in the session whenever the opportunity presents itself. Some portions of the orientation session (such as Block C) lend themselves more naturally to active involvement than others, but trainers should make an effort to model the kinds of teaching methods that they advocate at every opportunity. In this spirit, there is little point in monotonously lecturing new instructors to convey administrative or organizational information, except in circumstances where some kind of response or involvement from the new instructors is required (for example, signing people up for times in a mathematics "Help Room.") Whenever possible, this kind of information should simply be communicated in writing (e.g. via a memo or an E-mail) rather than verbally through a meeting. Such an approach has the advantage that training sessions may not be as boring for all participants, and the new instructors will also have a document to refer back to later in the semester. See the suggested readings for further suggestions on what kind of information may be appropriately communicated through writing.

Creating the Right Orientation Session for the Instructors You Have

An important point to recognize from the outset is that the training needs of the individuals who teach classes in your program will vary considerably from person to person. Perhaps the most tangible factors affecting the training needs are the amount of previous teaching experience, and the styles of teaching that each person has had experience with.

Lambert and Tice [70] describe in fairly general terms several training programs of note (see also the suggested reading section at the end of this chapter for other examples). Such programs often enjoy a large degree of homogeneity among their trainees. For example, the Duke University program focuses almost exclusively on preparing first year graduate students to become classroom instructors during their second year of graduate school. Unfortunately, such assumptions of trainee homogeneity are often not warranted. Departments employing a combination of graduate students, post-doctoral and new regular faculty, and adjunct faculty may find that there is a significant variation in the training needs of individuals. These departments should adapt the content, length, and focus of the orientation session accordingly.

In order to develop a convenient vocabulary, we distinguish among the following types of instructors.

Novice Instructors: Instructors with little or no experience in any kind of instructional role (neither as recitation or problem-session leaders nor as classroom instructors).

Journeyman Instructors: Instructors who have enough experience to feel confident in the teaching styles that they have specifically used, but who are not widely experienced in the use of the specific teaching methods described in this book.

Master Instructors: Instructors with wide experience of the specific teaching methods described in this book, or whose expertise and confidence permits them to easily learn and effectively employ new teaching methods.

In our experience, the compulsory inclusion of experienced teachers in orientation sessions designed for people who have never taught does little beyond breed resentment and bad feeling between the instructors and the training staff. An obvious but important exception to this guideline is the situation where master instructors are invited to participate in meetings as trainers or experts on teaching methods. Recognizing the existence and differing needs of these three groups, we have organized the orientation session into three blocks.

Block A: This block is intended for all instructors, and it includes introductions to the resources, traditions, and conventions of your mathematics department. For example, this block will cover typical methods of assessing students' performance in math courses, typical ways of calculating final grades, etc. This block may also feature the distribution of written handouts describing the details (photocopier access, office space, work expectations) of day-to-day life in your department.

Block B: This block is primarily intended for the novice and journeyman instructors. The intended purpose of this session is to serve as an introduction to some of the specific practices of student-centered instruction.

Block C: This block is primarily intended for novice instructors only, and it includes an opportunity to practice teaching and to receive feedback, as well as an opportunity to visit the classrooms of recognized master teachers early in the semester.

Naturally, this format can be adapted, if necessary, to better address the perceived needs of the trainees with which you will be working.

Other Assumptions About Instructors

In choosing the content for this orientation session (and for the manual as a whole) we have made two other assumptions about the instructors who will be trained. As a result we have not addressed these matters directly anywhere in this manual.

1. We assume that the instructor-trainees have sufficient knowledge of the mathematics that they will be expected to teach so that a lack of facility with the subject matter does not impair their teaching. We recognize that attempting to precisely define terms here is inherently problematic. We simply want to say clearly and up front that the present work does not include any measures to address problems with instructor knowledge of mathematics.

2. The sensitive issue of instructor facility with spoken English has received a great deal of attention in the media, and interest from students and parents at institutions of higher education. We simply wish to say clearly and up front that the present work does not attempt to identify or address potential problems in this area. Many institutions will also have an office or institute for monitoring, testing, and improving spoken English, and we feel that such institutions may be better equipped for handling such situations than the typical college mathematics department.

Goals for the Orientation Session

We have designed this preliminary orientation session to achieve a number of specific goals, which we have listed below. As you may expect, some of these goals are closely related to the teaching style and instructional environment that we envision. As a result, not all of the goals listed here may be readily applicable to your program. You should feel free to adapt these goals to better reflect the teaching and learning circumstances of your institution.

1. To provide new instructors with a clear articulation of the goals of the course that they are expected to teach.

2. To provide new instructors with examples of how other teachers have designed lessons to try to achieve course goals.

3. To introduce new instructors to the technology that they will be expected to use in their class.

4. To introduce new instructors to the grading systems and conventions typically used in the course that they are expected to teach.

5. To indicate to new instructors some of the issues and opportunities presented by the first day of class, and what kind of materials that they are expected to create for the first meeting with their class.

6. To familiarize instructors with the administrative duties that they are required to complete at the beginning of the semester.

7. To provide instructors with an introduction to the methods of student-centered instruction that they are likely to use on the first day or during the first week of class.

8. To provide instructors with some guidelines regarding the tests and quizzes usually given in your department and how these items are graded.

9. To provide instructors with an opportunity to practice organizing an explanation and presenting this explanation to others.

10. To provide instructors with an opportunity to receive feedback on their presentation skills.

11. To provide instructors with an opportunity to critically observe other teachers, and to consider how they may apply their observations to what they will do in their own classrooms.

12. To organize a program of visits by novice instructors to master teachers' classrooms early in the semester.

13. To provide training staff with some kind of preliminary indicator of the areas of instructor development on which the in-semester training should focus.

3.2   Agenda for the Orientation Session

1. Block A: An Introduction to Your Mathematics Course

    (a) Introduction of the instructor training team and mentors.

    (b) Overview of the training program.

    (c) Articulation of course goals.

    (d) Discussion of the difference between student learning and instructor presenting.

    (e) Distribution and discussion of sample lesson plans for the beginning period of the course.

    (f) Introduction to calculator and/or computer basics.

    (g) Advice and departmental requirements on designing a grading scheme.

    (h) Discussion of the important outcomes to accomplish on the first day of class.

    (i) Distribution of an administrative handout encompassing essential course information.

    (j) Opportunity for questions and answers.

2. Block B: An Introduction to Student-Centered Instruction

    (a) Discussion of the rudimentary issues of implementing cooperative learning in the classroom, such as how to form groups, how to get the students to work on the activities, how to wrap up the activities, and what kinds of activities to have students doing.

    (b) Brief outline of some of the difficult issues that can arise when using cooperative learning in the classroom.

    (c) Discussion of ways to encourage the students to read the textbook.

    (d) Discussion of the role that quizzes play in the course, and how to design and grade them.

    (e) Distribution of sample quizzes.

    (f) Discussion of issues of grading, such as the psychology of marking papers and how to grade fairly and efficiently.

    (g) Opportunity for practice grading.

3. Block C: Practice Teaching

    (a) Arrangement of a system of first-week visits to classrooms of experienced instructors.

    (b) Opportunity for practice teaching.

    (c) Feedback on practice teaching.

3.3   Block A: An Introduction to Your Mathematics Course

Goals for Meeting Block A

1. To provide instructors with an introduction to the training staff and an overview of the training program.

2. To encourage instructors to think about their teaching decisions in terms of the course student learning goals rather than their "performance."

3. To provide instructors with sample lesson plans to be used for the beginning period of the course, and to use these to give instructors an image of how a student-centered lesson plan may look.

4. To bring instructors to a competency in the basic implementation of the technology used in the course.

5. To assist instructors in designing an appropriate grading scheme.

6. To equip instructors to achieve important outcomes on the first day of class.

7. To convey all other course information essential for beginning the course.

8. To provide instructors with an opportunity to get their immediate questions answered.

Preparation for Meeting Block A

1. Identify all of the persons involved with the departmental training, as well as any other interested and experienced instructors, and determine each person's role in running this meeting.

2. Distribute to the instructors the essential course materials. The instructors should obtain a copy of the textbook, and a calculator or copy of the computer software needed for the course.

3. Distribute essential course information, such as when and where the courses will meet, when and where the training meetings will meet, and what office space will be given to each of the instructors.

4. If your department has not already done so, gather two or three experienced instructors to develop an official policy on grades that can be shared with instructors and students alike. This policy could be codified in a handout that includes an "official" departmental interpretation of each of the possible grades, any departmental grade requirements, historical course grade averages, and typical student reactions to the different grades.

5. Prepare a course handbook that codifies useful information for new instructors. See the outline for running this portion of the meeting for a list of possible items to include in the handbook. Many of these items will be referred to during Block A of this meeting, including

   • an outline of the training program,

   • an outline of the course goals,

   • a sample first day handout,

   • a sample student data sheet, and

   • information on grades.

6. Prepare an overhead outlining the training program.

7. Prepare an overhead articulating the goals of the course.

8. Prepare roughly one week's worth of lesson plans for the new instructors to use in teaching their course.

9. Identify the basic calculator or computer software skills needed by the instructors, and prepare a brief activity designed to encourage the instructors to explore these skills.

Outline for Running Meeting Block A

:01–:10 Welcome the trainees. Physically identify members of the training team and any other experienced instructors who are assisting with this meeting or with mentoring new instructors. Identify the roles that will be played by the members of the training team. If you have time, and if the trainees do not already know one another, have them each give brief introductions.

Distribute the course handbook. This handbook should include any information about the course, the department, the students, or the institution that the instructors may need to access during the semester. Items that may be included in this handbook are

   • an articulation of the departmental philosophies and course goals,

   • a description of the responsibilities of the training staff and of the classroom instructors,

   • an overview of the program of weekly meetings,

- a description of how the instructors' teaching will be evaluated and what this information will be used for,
- a description of the course textbook and the technology used in the course (emphasizing how these fit with the course goals),
- a sample first-day handout,
- a sample student data sheet for gathering information from the students,
- a semester-long day-to-day schedule suggesting what sections of the textbook to cover each day,
- recommended homework problems to assign from each section of the textbook,
- university and departmental expectations,
- information on departmental resources (e.g., computers, copiers, office supplies, and secretarial support),
- sample exams for the course from prior semesters,
- an overview of the student population,
- an overview of the departmental curriculum and how this course fits into it,
- necessary information on grades (e.g., departmental grade requirements, historical course grade averages, typical student interpretations of grades, and a sample grading scheme),
- a university calendar with important dates, and
- a description of any student support services available through the department or institution.

Refer the instructors to the part of the handbook that summarizes the training program. An overhead of this information would be helpful to have. Briefly talk them through the training program. Be sure to include a discussion of how the new instructors' classroom performances will be evaluated, and what this information will be used for. Briefly discuss how your professional development program fits into the larger context of national trends in undergraduate mathematics instruction and instructor professional development.

:10–:20 Articulate the goals of your course. Use an overhead, and refer the instructors to where the goals are summarized in the course handbook. Explain to the instructors why these particular goals are important. Situate the goals in terms of the national mathematics reform movement and in terms of the purposes of your particular institution. Describe briefly, in terms of current understandings of how students learn, why traditional forms of instruction may not be the most effective ways of realizing these goals.

:20–:35 Present to the instructors the oft-held traditional model of classroom instruction and learning as a process of encoding information (preparing the lecture), transmission of encoded information (giving the lecture), reception of encoded information (listening and taking notes), and decoding information (understanding the lecture). Ask the instructors to take a few minutes to discuss with one or two peers the following three questions.

1. Does this model describe my experiences as a college student?
2. Is this the model that I envision when I imagine my teaching?
3. How are the responsibilities for the success of student understanding under this model shared between the instructor and the students?

Lead a brief discussion in which you allow some of the instructors to share their responses with the whole group. Be sure to emphasize that under this model the instructor is almost wholly responsible for its success, other than rudimentary student responsibilities such as staying awake and paying attention.

Contrast this model with an outline of a classroom climate where students and instructors both view students as active processors of stimuli rather than passive receivers of information. Briefly discuss current theories of constructivist epistemology, and how through these the focus has been transferred from instructor performance to student understanding. Articulate some of the vocabulary used throughout this book, and elsewhere, in reference to student-centered instruction. Point out that it is important for the instructors to free themselves of the two assumptions

- that students can only possibly learn mathematics if it first proceeds out of the instructors mouth, and

- that students can only possibly attain a poor imitation of the mathematics that they see their instructor do.

If there is time, allow the instructors a few minutes to voice any questions or concerns about this model. Be sure to emphasize that the rest of this training program is meant to assist them in implementing a teaching style that is compatible with this latter view of classroom instruction and learning.

:35–:50 Hand out at least one week's worth of lesson plans for the new instructors to use to teach their classes. Ask the instructors to read one of the sample lesson plans. Then ask them to discuss with one or two peers their reactions to the plan. Lead a whole group discussion of the lesson plan, and explicitly tie the contents of the plan to the course goals and to the student-centered learning environment described previously. Summarize the important features of the sample lesson plans, and stress the importance of translating these into the forthcoming instructor-created lesson plans. These important features may include

- learning *activities* as part of the classroom,

- limits on the amount of lecturing,

- cooperative learning,

- use of calculators and technology to investigate mathematics,

- feedback and assessment items that are in harmony with course goals rather than working against course goals, and

- lessons planned under the assumption that the students have read the textbook prior to class.

:50–:70 Ask the instructors to get out their calculators. (If your program uses computer algebra systems or other computer software instead of calculators, modify the instructions in this section appropriately to introduce your trainees to the basics of your platform.) Lead a short presentation on the very basics of calculator usage, such as how to use the calculator's syntax for arithmetical calculations. Show the instructors the basics of graphing, leading up to using the calculator's graphical calculation abilities to solve problems, such as optimization. Ask the instructors to get with one or two peers, and give them a few short and basic exercises suitable for solving with a graphing calculator, and ask them to complete them at this time. For example, you could give them a typical optimization problem from a calculus course, ask them to set up a model of the problem, and then use their calculators rather than calculus to solve the exercise. Take a few minutes to allow the instructors to ask questions about the use of the calculators. Finally, alert all of the instructors, even the ones that feel comfortable using these machines, about the in-semester training meeting outlined in Chapter 11. Emphasize that this upcoming meeting will not be a re-hashing of the basics covered in this brief introduction, but a session on using technology meaningfully in support of course goals.

:70–:80 Give the instructor-trainees a ten minute break, if desired.

:80–:90 Make the instructors aware of information on any pre-determined common evaluations or other required common evaluation items in multi-section courses. Describe the most common evaluation techniques used in your department. In addition to the stalwart exams, quizzes, and homework, this may include team homework, group projects, writing, portfolios, presentations, and laboratory assignments. Pass out a sample grading scheme that weights these different assessment components to obtain a final grade. Tell the instructors about any "gateway" exams that your department uses. In order to give the instructors some context about the marks that they will be giving, refer them to the portion of the handout that is dedicated to grade issues. Briefly discuss the historical grade averages for the course, any institutionally published official meaning of grades, and typical student interpretation of grades at your institution. Finally, emphasize the importance of having a well-defined and well-articulated grading scheme to hand out on the first day of class that should not be modified after the fact.

:90–:100 Pass out a sample first-day handout, or syllabus, that details for the students the essential course information as well as instructor expectations. In addition pass out a sample student data form. One example is found in the "Meeting Materials" section of this chapter, and other examples are readily available [95]. The instructors can require their students to fill out a similar sheet on the first day of class. The information obtained from these data sheets can be used to help instructors get to know the students by name more quickly, to form cooperative learning groups, and to determine if there are any course placement problems. Summarize outcomes that are important to achieve on the first day of class in order to set the groundwork for a successful semester. These may include

- conveying all essential course information, requirements, and expectations;
- beginning to learn the names of the students;
- beginning the process of students learning each other's names;
- creating an atmosphere for active learning by providing students with opportunities to be active participants and active learners;
- getting the students into groups, if cooperative learning is to be a regular feature of the course;
- getting the students to read the textbook before the next class;
- assigning homework, if it is to be a regular feature of the course; and
- of course, doing some mathematics.

If anyone in your training staff or cohort of mentors has a clever activity for the first day of class, mathematical or otherwise, that works especially well in achieving one or more of the previously mentioned outcomes, have that person share it now.

:100–:120 Point the instructors to any key features of the course handbook that have not already been addressed during the meeting to this point. Allow the instructors some time to ask any pressing questions that they need answered before the first day of class. Make sure that those instructors who will be participating in Blocks B and C of this training session know where and when those will take place, and make sure that they know what they are to prepare for Block C.

3.4   Block B: An Introduction to Student-Centered Instruction

3.5   Goals for Meeting Block B

1. To equip instructors with the necessary strategies for initial implementation of in-class cooperative learning.

2. To alert instructors to possible difficulties involved in the implementation of in-class cooperative learning.

3. To provide instructors with initial strategies for encouraging the students to read the textbook.

4. To prepare instructors for designing a policy of quizzes that will meaningfully support course goals.

5. To prepare instructors for writing quizzes.

6. To assist instructors in articulating their expectations to students concerning grading of collected items.

7. To equip instructors to form initial homework teams, and to convey the procedures and roles to the students.

3.6 Preparation for Meeting Block B

1. Prepare a very brief activity that the instructors can do in groups while you model the in-class management of a cooperative learning activity. Since you will later discuss what types of mathematical problems are appropriate for such activities, you may choose a non-mathematical activity for the instructors to do here. In fact, that may be wise as to avoid undue competition among the instructors or distraction from the point of this meeting.

2. Prepare a list of suitable in-class cooperative learning activities that the instructors can use in the first few weeks of their courses.

3. Assemble a packet of sample quizzes generated by instructors in your department.

4. Assemble a couple of student responses to typical assessment items used by your department. These could be answers to homework problems, quiz problems, exam problems, etc. Another option would be to use the sample student response provided in the "Meeting Materials" section of this chapter.

3.7 Outline for Running Meeting Block B

:01–:05 Introduce the meeting, and thank the participants for attending. Briefly review the agenda for Block B. Make sure that the instructors are aware that Block B of this meeting is merely designed to equip them to successfully begin the semester, and that all of the items discussed in Block B of this meeting will be addressed in more detail in the weekly meetings throughout the semester.

:05–:30 Give a brief summary of the pros and cons of in-class cooperative learning. Some pros include the following.

- It allows the students to actively engage the mathematics.
- It allows the students to verbalize the concepts in both oral and written form.
- Groups of students can often successfully solve more difficult problems and engage more meaningful mathematics than can many individual students.
- Instructors can get immediate feedback on how well the students are understanding the material.
- Many students find it enjoyable.

The possible cons include the following.

- There are new classroom management issues that must be addressed.
- Some instructors are less comfortable with the amount of interpersonal interaction required to manage the activities.
- Care must be taken to ensure that all of the students are successfully engaging the material rather than passively watching group members do all of the work.
- Cooperative activities take a lot of time, so it may not be possible to "cover" as much material.

Model for the instructors how you would implement a cooperative learning activity in the classroom. Provide them with a "fake" problem to work on. Get the instructors into groups, and allow them a minute or two to work on the activity while you model walking around the room to interact with various groups. Call their attention to you, and wrap up the activity with a class discussion. Follow this role-play activity with a discussion of the nuts and bolts of using in-class cooperative learning, and give any relevant initial advice that you may have on these topics. Summarize that to implement in-class cooperative learning an instructor must be able to

- quickly get the students into groups,
- encourage them to start working on the activity rather than engaging in off-task discussions,

- interact with groups of students as they are working on the activity,
- get the students to stop what they are doing and pay attention to the front of the classroom at the appropriate time, and
- wrap up the activity as a class and draw out the meaning of the activity before going on to something else.

Hand out a list of cooperative learning activities that the instructors can use in their classrooms during the first few weeks of class. Briefly comment on why these activities were chosen as appropriate for in-class cooperative learning.

Finally, alert the instructor to difficulties that are commonplace in the implementation of in-class group work, and remind them that a later training meeting will help equip them with strategies for dealing with many of these situations. The potential problems include the following issues.

- Some will be passive and reluctant to participate.
- Some students may try to work on homework or other work, instead of the activity that you assign.
- Some students may be inclined to start off–task conversations, disrupting the learning of other students, or distracting you.
- Some students will be inclined to dominate their groups by attempting to do all of the work themselves and dismissing the contributions of others.

It is important to note that if in-class cooperative learning is implemented well, most students will be won over once they get used to it, because they will see the benefits in terms of increased learning and enjoyable classroom experiences.

:30–:45 Pass out the handout "Getting Students to Read the Book." Relate the handout to the previous discussion. Make sure that the instructors know that they must clearly communicate early in the semester, verbally *and through their behavior*, that they expect the students to read the textbook. The instructors should communicate to the students, verbally or in the syllabus, how reading fits into the goals of the course. Finally, you cannot emphasize too strongly the point that covering all the material in lecture as if the students have never read it is a strong disincentive for the students to continue reading the book.

Remind the instructors that lecturing is still one useful weapon in their arsenal. Well-placed and well-conceived short lectures can serve many functions, several of which relate to encouraging students to read the book. Some such appropriate uses of lectures include

- previewing the high points or difficult parts of the upcoming reading to enable students to better read the section on their own,
- providing an overview of previous reading including connections among concepts, and
- working through the hardest parts of a particularly difficult or important example from the text.

In all of these cases, it is helpful to use phrases like "as you saw in your reading last night," "as you will find when you read this section," etc.

Finally, reiterate that they should not plan lessons that encourage (even unintentionally) students not to read the book, and also reiterate the importance of the instructors reading the textbook themselves.

:45–:70 Ask the instructors to discuss for a couple of minutes with a peer what roles they think that quizzes will play in their courses. Lead a brief discussion with the entire group on the possible uses of quizzes, and how quizzes fit into the goals of the course. It is important for each instructor to articulate his or her reasons for giving quizzes, because these will affect how the instructor will design quizzes. Some possible goals of giving quizzes include

- to encourage attendance,

- to check if the students have done their homework and reading,
- to test students for basic competence with key skills,
- to test problem solving,
- to communicate the expectations of the instructor in advance of exams, and
- to provide regular feedback for both instructor and students on the students' progress.

Lead a group discussion centered on any of the following additional questions concerning quizzes that you deem important. It would be appropriate to give any guidelines or advice at this point.

- Should the questions be new to the students or ones that they've seen before?
- How long should quizzes be and how many points should each quiz be worth?
- How often should quizzes be given?
- Should the quizzes be announced or unannounced?
- How tough should the quizzes be? Notice that the answer to this question will likely depend upon the prevailing morale of individual classes. Sometimes it may be appropriate to toughen them up, and at other times it may be more motivating to make them slightly easier.

Point out that no matter what roles quizzes play in their courses, it is important to always give useful feedback when grading and returning the quizzes. More will be said about grading later.

Pass out the packet of sample quizzes that you have assembled from members of your department, and give the instructors a few minutes to look over them and ask any questions that they may have at this point.

:70–:80 Give the instructor-trainees a ten minute break, if desired.

:80–:100 Point out to the instructors that they will likely be getting queries very early on in the semester concerning their expectations on graded items, even before such items are to be collected. It is important that the instructors begin to think about their expectations, and how to communicate them to the students. These expectations may also play into the development of their overall grading scheme, which was addressed in Block A of this meeting.

In order to provide some context for this discussion, give the instructors copies of a couple of student responses to typical assessment items used by your department. Another option is to give them copies of the sample student answers provided in the "Meeting Materials" section of this chapter. Ask the instructors to take a couple of minutes to grade these. After a couple of minutes, ask the instructors to discuss with one or two peers how they graded the items, and their rationale for their choices. Lead a discussion with the whole group.

Point out that some difficult decisions include how fussy to be in terms of neatness, presentation, arithmetic, and grammar; how much to interpret or guess at the student's intentions when grading; and how to develop a reasonable yet firm late policy. It is also helpful to point out the importance of developing and using rubrics when grading papers. (See "Grading Rubrics" in Chapter 6.) These are useful for when students inquire about grading, for when students ask to have something re-graded, or for when you accept late papers that you want to grade in a consistent manner with the previously collected papers.

Point out that the marks they use when grading have psychological effects, so it is important to grade conscientiously. At this point you could include some advice about grading. This may include (as basics)

- Don't nit-pick. Correct all errors, but don't subtract points (or worse partial points) for every little error. Instead, give a number grade to each exercise (or partial exercise).
- Give a score to each exercise, rather than taking away points from each exercise. In other words, if a problem is scored out of ten points, instead of writing "–3," write "7."

• Do not use emphatic markings when grading, especially when correcting mistakes. These may be misinterpreted and consequently make students feel demeaned.

More detailed advice on grading can be found in Chapter 6, although it is probably not necessary to include it all now since grading will be addressed in more detail in that meeting.

:100–:120 Lead a discussion on the rationales for using homework teams. Homework teams have many of the same rationales as in-class cooperative learning activities. In addition, they can provide a network of peers available for consultation. They can also provide a system of accountability. Finally, they give the instructor the flexibility to require students to submit written reports on extensive problems or projects that would be prohibitive for the students to do alone, as well as prohibitive for the instructor to grade, if each student were to do them individually.

Pass out the handout entitled "Roles in Homework Teams." Use this to explain the various roles played by members of the homework teams on each assignment. Discuss the departmental policies on reporting roles played, and rotating roles throughout the semester. Other possible breakdowns of roles played within homework teams can be found in the MAA Notes series, for example [43] and [99].

Emphasize the importance of communicating to each class the key points of the previous discussions about homework teams. The students should know that the homework teams are being used for well-intentioned purposes, and that these purposes coincide with the goals of the course. The students should also be made explicitly aware of the expectations concerning roles, and they will need to be assisted in adequately performing these roles. It is helpful to express that, aside from various personal issues that arise with varying frequency, many students find that they enjoy the experience of working on homework in teams outside of class. In addition to this, more and more students are coming into college mathematics courses having already experienced working in teams in high school or in other college courses.

Provide the instructors with a recommended framework for assigning initial homework teams. One possibility is found in the handout "Forming Initial Homework Teams" in the "Meeting Materials" section of this chapter. If homework teams will not be formed until a few weeks into the semester, there are more possibilities for the ways of forming the initial homework teams. If this is the case in your department, you may consult the handout entitled "Forming Homework Groups" in Chapter 7 of this book. Convey any departmental policies concerning how often to change homework teams throughout the semester. If you do not already have a departmental policy, some useful resources for guiding your decision include works in the MAA Notes Series, for example [95] and [99]. Some practitioners hold strongly to the belief that semester-long groups are the most beneficial, while others believe that it is important to re-form the groups periodically. In truth there are certain advantages to each system that must be considered in the context of your institution.

Discuss typical student and instructor reactions to homework teams at your institution. Give the instructors a few minutes to express any questions that they have, or reservations that they may be feeling, about using homework teams. Do not feel that you need to provide an answer to every reservation that instructors express. Some people will express doubts over very robust and thoroughly tested plans simply because they are not personally comfortable with the thought of working with or depending upon others.

3.8   Block C: Practice Teaching

As indicated, this block is intended primarily for the novice instructors. This block is longer than the other two, but many novice instructors feel that the extra time is well spent, as this block can provide them with an experience that is closer to the actual act of teaching than most of the discussion sessions that the orientation thus far has provided them.

Goals for Block C

1. To provide instructors with an opportunity to practice organizing an explanation and presenting this explanation to others.

2. To provide instructors with an opportunity to receive feedback on their presentation skills.

3. To provide instructors with an opportunity to critically observe other teachers, and to consider how they may apply their observations to what they will do in their own classrooms.

4. To organize a program of visits by novice instructors to master teachers' classrooms early in the semester.

5. To provide training staff with some kind of preliminary indications of what areas of instructor development the in-semester training should focus on.

Preparation for Block C

1. Find out when master teachers will be teaching their classes, and contact the master teachers to make sure that they are comfortable with the idea of being visited. In our experience, once you explain the purpose of the visits, most master teachers will be very happy to cooperate.

2. Arrange for a suitable room that is large enough to accommodate all of the participants and the video equipment. If you have a lot of people who wish to participate, it may be advisable to arrange for several rooms and then break up into smaller groups of four to six instructors and one trainer to each room.

3. Make sure that the rooms are equipped with the presentation devices that instructors in your course normally use. At a minimum, this will normally be a chalkboard and some chalk, but it may also include overhead projectors, colored chalk and colored pens, graphing calculators with overhead projector viewscreens, computers with projectors, etc.

4. Make sure that each room has the equipment necessary for videotaping and watching the instructors. This will include the following items.

   (a) A video camera with good microphone and tripod

   (b) A video cassette player and television monitor

   (c) A video cassette

   (d) A stop-watch or timer with an audible alarm

   (e) Copies of the "Guide for Evaluating Explanations" found in the Meeting Materials section of this chapter

   (f) Name tags and writing materials

   (g) (Optional) Refreshments

5. Ensure that you are thoroughly familiar with the operation of the video equipment that you have, and that you can explain its operation quickly and concisely to others.

6. If you plan to have the instructors solve and explain a mathematical problem, then prepare some problems to distribute to the instructors. In any case, it can be helpful to develop a list of fundamental topics or concepts from the course that the instructors will be teaching, to help them settle on a subject for the lesson that they will develop. In order to avoid monotonous repetition, each instructor should give a lesson on a different topic or solve and explain a different problem. Try to find topics, concepts, and problems that can be dealt with reasonably well in five minutes if the person knows what they are doing.

7. (If this block is separated from Block B by a break.) Make sure that all participants know what room they will be meeting in, and what time they are expected to arrive.

An Outline for Running Block C

:00–:05 Welcome the participants to the session. Briefly explain how the session will be run:

1. Each instructor develops a lesson.
2. Each gives the lesson and is videotaped.
3. Everyone watches the videotapes and offers suggestions.

Stress that the instructors should strive to develop a mini-lesson that is no more than five minutes long. Distribute the topics or problems to the instructors and ask them to spend the next fifteen or twenty minutes developing their mini-lessons or problem solutions and explanations. Stress to the instructors that they should attempt to explain at a level that they think is appropriate for the undergraduate students that they will have to teach. Also, point out the different items of equipment in the room (overheads, colored chalk, calculators) that instructors may use in their lessons, if they so choose.

:05–:25 Allow the instructors time to think about and develop their lessons. It may help to let the instructors know that you are very happy to assist with the production of visual aids, etc., and that, if the instructors are unsure about whether their lesson is at the right level, they can ask you as well. It is important for you to resist the temptation to add too many of your ideas to the instructors' work, though, as they may just end up letting you plan the lesson for them.

:25–:30 If you plan to have instructors operating the video camera to videotape each others' lessons, then spend a few minutes showing the instructors the basics of the particular video camera that you have. They should know how to start and stop recording, how to zoom, and how to pan.

:30–:65 Have each of the instructors give their lesson one after the other. The time allotted here is based on an assumption that you will have five instructors, each of whom will take about five or six minutes, and there will be about a minute of "down time" between each lesson. Have the instructors use the stopwatch to time themselves, and insist that they end quite quickly as soon as their time is up. When they finish, some instructors will be so focused on how their lesson went that it will be impossible for them to think about anything else. To alleviate this, it can be useful to have the instructor who has just finished giving their lesson videotape the next person. When each person finishes their lesson, be sure to thank them for their contribution.

:65–:70 If you choose to, it can be useful to have a short break here. This will also give you a chance to rewind the videotape and to set up the VCR, television monitor, and chairs for viewing the videotape.

:70–:80 Reconvene the instructors and explain how the next part of the session will work. Tell the instructors that they will all watch the video clips one at a time, using the "Guide for Analyzing Explanations" (see the "Meeting Materials" section of this chapter) to note whatever they see in the clip that they think is effective, and any suggestions they have for how the instructor could be even more effective in the future. Tell the instructors that when they come to give feedback, the group will begin with the positive comments, and then move on to the suggestions.

:80–:155 Carry out the program described above. The time budget here is five instructors at fifteen minutes each. This fifteen minutes may seem excessive, but when broken down, there is relatively little time.

5 or 6 minutes: Watching the video clip.

1 or 2 minutes: An opportunity for instructors to jot down last minute notes after watching the clip.

3 or 4 minutes: An opportunity for the instructor who was on the video clip to say what he or she saw that seemed to be effective.

6 or 7 minutes: Comments from other instructors.

The trainer who is leading the discussion will probably have to manage time quite carefully during this part of the session. Some instructors have a tendency to make rather long statements about the video clip, sometimes with no clear point or suggestion. It can be helpful for all concerned, and especially the instructor who was on the video clip, to gently focus the discussion to practical issues and concrete suggestions. We have found that it is usually helpful to give the instructor who was on the video clip the opportunity to speak first. However, some instructors have a tendency to immediately focus on what they saw in the clip as "wrong," especially in their own teaching. While it is certainly important to get to these issues eventually, it is also important to note what was effective. This is an area where the trainer leading the meeting may also have to gently insist that the instructors begin with what they saw as effective, before moving on to suggestions. It is worthwhile for the trainer to note some effective points in the lesson and be prepared to contribute these to get the discussion going. For further suggestions on stimulating discussion, see Appendix A. It may also be helpful to review the procedure for conducting the follow-up meetings to observational class visits found in Chapter 17, especially the characteristics of constructive feedback.

:155–:170 This time is available in case fifteen minutes is not sufficient for each of the instructors, or if you have more than five instructors. If the session has stayed on schedule, then this time will be unnecessary and you should proceed directly to the wrap–up.

:170–:180 Arrange for the instructors to find a time to visit a master teacher's classroom during the first week of class, if possible. You should tell the instructors exactly where and when the class meets, especially if they are new to your institution and don't know their way around. You should also arrange for some kind of feedback or reporting mechanism. For example, instructors could use the "Guide for Analyzing Explanations" while they are observing the master teacher, and then turn a copy of this in to you. End the meeting very positively. Remember, the next time that these instructors will be in front of a group of people will probably be the first meeting with their class. These people will need as much confidence as you can instill in them.

3.9   Evaluating the Orientation Session

Most of the training meetings described in this book suggest the use of a questionnaire that appears in Appendix A as "Meeting Evaluation Questionnaire." In the "Meeting Materials" section of this chapter, we include a different questionnaire for evaluating this orientation session that reflects the differences in organizational structure and content of the orientation session as compared to the in-semester training meetings. As with all other suggestions offered here, you should feel free to use this questionnaire as a starting point for developing your own.

3.10   Meeting Materials

This list includes all of the meeting materials required for the orientation. Items **1** to **6** are intended for use during **Block A**, items **7** to **13** are for **Block B**, and item **14** is intended for use during the practice teaching session, **Block C**. Item **15** is the evaluation form, and should be distributed to all instructors who attended the pre–semester orientation.

1. Copies of a course handbook that includes all essential information that the instructors will need to have at hand throughout the semester.

2. An overhead outlining this training session and the entire training program.

3. An overhead outlining course goals.

4. Copies of lesson plans for the new instructors to use during at least the first week of the course.

5. Copies of sample first-day handouts or syllabi.

6. Copies of the handout entitled "Student Data Form." (See sample.)

7. Copies of an activity to be used to introduce the instructors to in-class cooperative learning.

8. Copies of suggested in-class cooperative learning exercises for the instructors to use during the first few weeks of the course.

9. Copies of the handout entitled "Getting Students to Read the Book."

10. Copies of a packet of sample quizzes from instructors in your department.

11. Photocopies of different "Student Responses to a Sample Problem" for grading. (See the examples provided.)

12. Copies of the handout entitled "Roles in Homework Teams."

13. Copies of the handout entitled "Forming Initial Homework Teams."

14. Copies of the handout entitled "Guide for Analyzing Explanations."

15. Copies of the questionnaire entitled "Mathematics Teaching Orientation Session Evaluation Form."

STUDENT DATA FORM

1. Your Name (and what you would like to be called in class):

2. Campus Address:

3. Phone Number:

4. Year in School (e.g., Freshman, Sophomore, etc.):

5. Planned Major:

6. What is the most advanced math course you have taken?

7. Why are you taking this course? (Please write at least two sentences.)

8. What grades do you usually get in math courses?

9. How do you feel about math? (Please write at least two sentences.)

10. How do you feel about the prospect of working in groups? (Please write at least two sentences.)

11. When would be the best time for you to meet your group: afternoon, evening, or late night? (List first and second choices.)

12. What other courses are you taking this semester?

Getting Students to Read the Book

Bob Megginson[1]

1. Don't lecture as if the students have never before seen the material!

2. Don't lecture as if the students have never before seen the material!

3. Don't lecture as if the students have never before seen the material!

4. You must *really* expect them to read the book, and always act as if you expect them to read the book.

5. You must set the tone the first day of class, *saying* that they must read the book and why.

6. *You* must read the book!

7. Make each day's assignment of reading an *event*, complete with coming attractions.

8. When you start the day's activities, do a brief activity that assumes the reading of the book.

9. As a rule, *don't* do examples directly from the book, unmodified—it sends the wrong message. (But there are exceptions.)

10. When examples from the book are important and difficult, go over the *difficult* parts, only outlining the *results* of the easier parts, constantly tossing in phrases such as "as you saw in your reading...." Better yet, get them involved in a group activity to work through the exercise themselves.

11. When they *aren't* doing the reading, try:

    (a) Brief quizzes on the reading at the beginning of each day.
    (b) Group activities based on the reading, as mentioned above.

Most importantly–

12. Don't lecture as if the students have never before seen the material!

---

[1]Used with permission. This handout has also appeared in [103].

Student Responses to a Sample Problem

2. (1 point) Does the sequence, $\qquad c_n = \dfrac{\sin(n)}{n}$

converge or diverge? Explain why you think your answer is appropriate.

I would think it does both .. The graph of the function $C_n = \dfrac{\sin(n)}{\wedge}$ look like this:

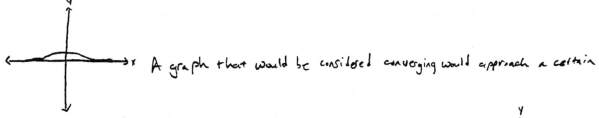

A graph that would be considered converging would approach a certain value. Ex: ___ and a diverging moves away from a value. Ex. diverging.

This function moves away (diverges) from 0 and the approaches (converges) on it, so it is neither exclusively converging or diverging, but does both.

2. (1 point) Does the sequence, $\qquad c_n = \dfrac{\sin(n)}{n}$

converge or diverge? Explain why you think your answer is appropriate.

since the sin (n) function alternates between increasing and decreasing and positive and negative the sequence $\dfrac{\sin(n)}{n}$ does not converge, however it does. remain near 1 at all times

2.   (1 point) Does the sequence, $c_n = \dfrac{\sin(n)}{n}$

converge or diverge? Explain why you think your answer is appropriate.

$$c_n = \frac{\sin(1)}{1} \; , \; \frac{\sin(2)}{2} \; , \; \frac{\sin(3)}{3} \; , \; \frac{\sin(4)}{4} \; , \; \dots \; , \; \frac{\sin(n)}{n}$$

n	c_n
1	.8415
4	-.189206
10	-.054402
20	.04545
25	-.00529
40	.01843
100	-.0506

The values of c_n seem to be randomly oscillating between very small negative and positive numbers. However, the sequence (values of c_n) do not approach a limiting values as n tends to infinity. Therefore, the sequence DIVERGES.

2.   (1 point) Does the sequence, $c_n = \dfrac{\sin(n)}{n}$

converge or diverge? Explain why you think your answer is appropriate.

Converges - As n increases, the denominator increases. The numerator is always between -1 and 1, and the absolute value of c_n decreases as $n \to \infty$. Therefore, c_n approaches zero as $n \to \infty$

Roles in Homework Teams

On each homework set, the members of your team will play certain roles. These roles should be rotated so that everyone shares different responsibilities for the team's success. The roles are the scribe, the clarifier, the reporter, and the manager.

Scribe: The scribe is responsible for writing up the single final version of the homework to be handed in. This is the set of solutions that will be graded. Each member of the team will receive the same grade. Whenever possible, your solutions should include symbolic, graphical, and verbal explanations or interpretations. Diagrams and pictures should also be provided whenever possible. The homework should be neatly written or typed, and stapled or paper-clipped.

Clarifier: During the meeting, the clarifier is responsible for assisting the team by paraphrasing the ideas presented by the other team members. The clarifier is responsible for making sure everyone in the team understands the solutions to the problems before the team moves on. You are working together so that all may better understand. Thus, the role of the clarifier is the most important, but sadly the most neglected, role.

Reporter: The reporter is responsible for writing a record of how the homework sessions went; where, when, and how long the team met; what difficulties or successes the team may have had (with math or otherwise); and so forth. The report should list the members of the team who attended the sessions and their roles. If a team member was absent from one or more meetings, this should be noted on the report. The report should be on a separate sheet of paper, and it should be the first page of the homework.

Manager: The manager is responsible for arranging and running the meetings. The manager is to ensure that all members know where and when the meeting will take place. If willing, the manager should arrange for refreshments for the meeting. If the team only has three members for a meeting, the manager should take up the role of the missing person. When the homework is returned, the manager should see that it is photocopied and distributed to each team member.

Forming Initial Homework Teams

Here is one method of forming initial homework teams for the semester.

1. Gather all of the completed student data forms.

2. Collate all of the data forms into three piles according to the first preference for team meeting time.

3. Put students within each pile into groups of four according to the proximity of their location on campus.

4. As far as is possible, try to make these groups heterogeneous according to the students' gender, year, major, mathematical experiences, mathematical ability, and attitudes towards mathematics.

5. Move some students from the pile with their first preference of meeting time to the pile with their second preference of meeting time if necessary.

Guide for Analyzing Explanations

Material and content

1. Instructor states the topic clearly at the beginning. ____

2. The explanation is organized. ____

3. Instructor emphasizes the main idea. ____

4. Instructor relates material to something students already know. ____

5. Instructor gives suitable examples. ____

6. The material is appropriate to the level of beginning undergraduates. ____

7. Instructor has a clear conclusion. ____

Technique and delivery

8. Instructor involves students. ____

9. The instructor shows enthusiasm. ____

10. Board work is clearly organized and readable. ____

11. The pace is appropriate. ____

Comments

Mathematics Teaching Orientation Session Evaluation Form
Date: / /

1. Which term do you feel best describes your level of teaching experience and familiarity with the teaching techniques emphasized in this training program (please circle one):

   (a) beginner

   (b) some (at least one semester) experience

   (c) very experienced

2. Please circle the sessions of the training program that you attended.

   (a) Block A: An Introduction to the Course

   (b) Block B: An Introduction to Student-Centered Instruction

   (c) Block C: Practice Teaching

   Please circle your response to each question (Strongly agree to Strongly disagree) for each of the blocks that you attended.

3. Block A: An Introduction to the Course

   (a) The goals of the introductory course were clearly described.
   (Strongly Agree   Agree   Neutral   Disagree   Strongly Disagree).

   (b) Differences between a focus on student learning and a focus on presenting material were described in a way that I could understand.
   (Strongly Agree   Agree   Neutral   Disagree   Strongly Disagree).

   (c) I have a clear idea of some of the ways in which lesson planning can contribute to course goals.
   (Strongly Agree   Agree   Neutral   Disagree   Strongly Disagree).

   (d) I feel that I can competently perform basic operations with the graphing calculator.
   (Strongly Agree   Agree   Neutral   Disagree   Strongly Disagree).

   (e) I feel that I can design a grading system, and I know where to turn for help if I get stuck.
   (Strongly Agree   Agree   Neutral   Disagree   Strongly Disagree).

   (f) I feel that I know what documents to prepare for the first day of class, and I have some ideas for the kinds of classroom activities that I can have students perform on the first day of class.
   (Strongly Agree   Agree   Neutral   Disagree   Strongly Disagree).

   (g) The session stayed on topic and on schedule.
   (Strongly Agree   Agree   Neutral   Disagree   Strongly Disagree).

   (h) *Please use the space provided below for additional comments on Block A.*

4. Block B: An Introduction to Student-Centered Instruction

   (a) I have a procedure for grouping students during class.
     (Strongly Agree   Agree   Neutral   Disagree   Strongly Disagree).

   (b) I feel that I know what management functions that I have to perform during cooperative learning.
     (Strongly Agree   Agree   Neutral   Disagree   Strongly Disagree).

   (c) I am aware of some of the difficulties that may arise when cooperative learning groups are used in class.
     (Strongly Agree   Agree   Neutral   Disagree   Strongly Disagree).

   (d) I recognize the importance of teaching in a way that encourages students to read the text book.
     (Strongly Agree   Agree   Neutral   Disagree   Strongly Disagree).

   (e) I have thought about the goals that I will use quizzes to try to achieve in the course.
     (Strongly Agree   Agree   Neutral   Disagree   Strongly Disagree).

   (f) I am aware of the length and level of difficulty that is typical for quizzes given in this department.
     (Strongly Agree   Agree   Neutral   Disagree   Strongly Disagree).

   (g) I recognize that quizzes can provide feedback to both me and the students on the students' learning.
     (Strongly Agree   Agree   Neutral   Disagree   Strongly Disagree).

   (h) I have a procedure for grouping students into homework teams.
     (Strongly Agree   Agree   Neutral   Disagree   Strongly Disagree).

   (i) I can adequately articulate to my students how to function in the roles required for the homework teams.
     (Strongly Agree   Agree   Neutral   Disagree   Strongly Disagree).

   (j) The session stayed on topic and on schedule.
     (Strongly Agree   Agree   Neutral   Disagree   Strongly Disagree).

   (k) *Please use the space provided below for additional comments on Block B.*

5. Block C: Practice Teaching

   (a) The opportunity to prepare and present a piece of mathematics was helpful.
     (Strongly Agree   Agree   Neutral   Disagree   Strongly Disagree).

   (b) The opportunity to watch myself teaching was helpful.
     (Strongly Agree   Agree   Neutral   Disagree   Strongly Disagree).

    (c) The suggestions provided were useful.
        (Strongly Agree   Agree   Neutral   Disagree   Strongly Disagree).

    (d) Watching and analyzing other instructors' teaching gave me ideas for how to improve my own teaching.
        (Strongly Agree   Agree   Neutral   Disagree   Strongly Disagree).

    (e) The session stayed on topic and on schedule.
        (Strongly Agree   Agree   Neutral   Disagree   Strongly Disagree).

    (f) *Please use the space provided below for additional comments on Block C.*

6. *Please use the space provided below for additional comments on the orientation session as a whole. We are particularly interested in any practical suggestions that you have for how the sessions could be improved in the future. Thank you!*

3.11   Suggested Reading

- Pat Shure, Beverly Black, and Doug Shaw. *"The Michigan Calculus Program Instructor Training Materials."* [103]

  This document describes a five day, intensive pre-semester training program in considerable detail. The book includes a schedule for all of the training sessions that it describes, examples of the kinds of documents used and distributed to instructors, and detailed instructions for running each training session. This is the pre-semester training program that we feel is most compatible with the in-semester training meetings and classroom visits that we describe in this book.

- Bettye Anne Case. *"Responses to the Challenge: Keys to Improved Instruction. MAA Notes #11."* [23]

  The first part of this report contains views from members of the 1987 MAA committee on Teaching Assistants and Part-time Instructors. Many of these views include information about the training needs of new instructors (especially novices) and offer some suggestions of the kinds of training activities that can be provided for new instructors. The report also includes a chapter on foreign teaching assistants, English as a second language, language tests, and resources. The appendix to this report includes a number of "Teaching Handbooks" from mathematics departments, manuals for foreign teaching assistants, examples of end-of-semester evaluation forms, and some descriptions of training programs in mathematics departments.

- Leo Lambert and Stacey Lane Tice, eds. *"Preparing Graduate Students to Teach."* [70]

  This volume describes the content and organization of a number of training programs in several disciplines. Because of space constraints, the descriptions are somewhat abbreviated. However, there is still quite a lot of information about the kinds of training activities that can be provided for new instructors. The volume also includes a discussion of some of the problems encountered in training new instructors, especially graduate student instructors.

- Ron Douglas, ed. *"Toward a Lean and Lively Calculus. MAA Notes #6."* [42]

  Although this volume provides no direct advice for designing or implementing an intensive pre-semester training program, we include it here because it contains a wealth of opinions regarding the issues involved in teaching introductory college mathematics courses such as calculus. The volume can serve as an excellent starting point for departments that are interested in clarifying and codifying the goals of their introductory programs, and exploring the teaching and learning issues that the goals of the introductory programs may raise. For example, one goal may be to increase students' abilities to solve non-routine problems, and the department would need to identify teaching methods that would assist in the realization of this goal.

- Alan Tucker and James Leitzel, eds. *"Assessing Calculus Reform Efforts: A Report to the Community. MAA Reports #6."* [115]

  As with the previous reading, this report does not make direct recommendations regarding the training of mathematics instructors. However, it does report on a wide variety of reform-oriented introductory programs, such as Project CALC, C^4L, Calculus and Mathematica, etc. The report includes statements of many of the goals and some of the first experiences of the people implementing these programs. Like the previous reading, this report may serve as a starting point for those wishing to clarify the goals of their introductory programs.

- Carl Rinne. *"Excellent Classroom Management."* [97]

  Chapter 6 of this book discusses some of the opportunities that the first day of class presents, and the importance of making the right sort of impression on students on the first day. The discussion stresses the importance of impressing upon students "that your classroom is all about *interesting learning* (page 75)."

- Pat Shure, Morton Brown and Beverly Black. *"Michigan Introductory Program Instructor's Guide."* [9]

This handout is routinely provided to new instructors in the Michigan Introductory Program. Much of the content of this document supports the pre-semester training program described in [103]. The table of contents of this document is included below.

1. Welcome to the Michigan Introductory Program.
2. The Spirit of the Introductory Courses.
3. Organization of the Introductory Course System.
4. Grades and Grading.
5. Components of the Course.
6. Classroom Organization and Management.
7. The Instructor-Student Relationship.
8. University Tutoring Facilities.
9. University Support Offices: Academic and Personal.

- Michael Reed and Lewis Blake. *"Instructor's Manual for Math 31L and Math 32L."* [93]

This handout is routinely provided to new instructors in the introductory Laboratory Calculus program at Duke University. The idea of this document is that much of the administrative information that instructors need to know in order to run their course is best provided as a written document that instructors can refer back to when necessary, rather than in a meeting. In a meeting, instructors may be less likely to pay attention, and consequently they may have difficulty recalling the information later. The table of contents of this document is listed below.

1. Outline of *Crucial Information*. (This includes: texts, syllabus, labs, calculators, tests, homework, honor code, final exam and grading.)
2. Syllabus and Homework.
3. Lab Information, Policies and Choices.
4. Group Work and Individual Responsibility.
5. Tests.
6. Grading Policies.
7. Supplies, Aids, Miscellaneous Information.

- Gary Althern. "A Manual for Foreign Teaching Assistants" in [23]

This short booklet is also available directly from the University of Iowa. This booklet describes some of the ways in which the behavior and assumptions of American undergraduates may be misinterpreted by foreigners. The booklet also suggests areas of language usage that may confuse foreigners, both in terms of the meaning conveyed by the language and the respect (or perhaps perceived lack of respect) conveyed by the choice of language. The booklet includes suggestions for helping foreign teaching assistants to interpret and respond to their students.

- Committee on the Teaching of Undergraduate Mathematics, Mathematical Association of America. *"College Mathematics: Suggestions on How to Teach It."* [85]

This booklet is over twenty years old, but it contains suggestions for instructor behaviors that are still relevant today. The authors acknowledge that the booklet concentrates exclusively on the lecture method of instruction. However, much of the discussion of the issues that can be important on the first day of class, and the advice for encouraging instructor-student interaction during the class, are very relevant to those using student-centered instructional methods. Sections of this booklet may be particularly useful to beginning instructors, as the information is organized into itemized lists that use very direct language. On the whole the presentation is quite unequivocal (although the authors of the booklet are careful to point out that they are not trying to give prescriptions, simply suggestions using unequivocal language), which may make them more attractive to beginners.

Chapter 4

Making In-class Groups Work

Since the Tulane conference [42], substantial amounts of time and resources have been put towards developing materials for non-lecturing, student-centered classroom techniques. Some techniques that are used by many instructors are cooperative learning activities, student presentations, laboratory activities, discovery learning, and the Moore method. In our classrooms we extensively use the technique of having students work in cooperative groups on mathematical problems. In this chapter we focus on developing the instructors' facility with implementing and managing in-class cooperative activities. If your program includes student-centered teaching methodologies not addressed by one of the meetings in this book, Chapter 16 provides support for developing new training meetings.

Cooperative learning is a popular pedagogical device in elementary collegiate mathematics classrooms. There are many reports that suggest that when cooperative learning is implemented effectively in a mathematics class, there can be dramatic gains in what students are able to accomplish in terms of solving non-routine problems, in students' appreciation of and enjoyment in mathematics, and in students' willingness to persist in their attempts to solve and develop meaningful understanding of mathematical concepts that go beyond simple imitation of "worked examples" from the textbook. Although some investigators have reported some gains in terms of students' scores on standardized tests, the reports are not usually of dramatic gains [35]. A typical report is of the form "The students in the cooperative learning group performed at least as well as students in a control group." Because of the positive effects of cooperative learning that we have observed in our own teaching, we implement cooperative activities both in and out of class.

This meeting focuses on the use of cooperative groups in class. The use of cooperative groups in assigned activities outside of class is discussed in Chapter 7. In our teaching we use cooperative activities interspersed among short lectures, which are no longer than ten minutes in duration. These activities are intended to deepen students' understanding of the material of the lectures. The results of the activities are often reported and explained to the rest of the class by students, and the students also participate heavily in the establishment of the mathematical validity of the results presented. We see our role in this teaching method as that of facilitator and guide for student learning.

There are two paramount issues inherent in this teaching methodology: choosing appropriate group activities and successfully managing the implementation of these activities.

We see two overarching aspects to consider when choosing activities to be done in cooperative groups. First, the problems chosen should really require and encourage the students to work on them in groups. In other words, there should be a discernible difference between problems appropriate for individuals and problems appropriate for groups. The second aspect to consider is how the chosen problems facilitate the particular learning goals that the instructor has for that portion of the lesson. We begin this meeting with an analysis of some typical mathematics problems, and a discussion on their appropriateness for cooperative activities.

Even if cooperative learning can be shown with certainty to enhance student learning, implementation will remain a vital issue. Unfortunately, the positive results referred to above are mitigated by reports, also largely anecdotal, of disasters with cooperative learning, where the use of groups in a mathematics class appears to have produced very negative results. Again, the most decisive information seems to be mainly in the areas of students' attitudes towards mathematics, rather than in their achievement on standardized tests. Recognizing this, a literature has developed that is primarily concerned with describing methods for conducting cooperative learning, complete with concrete suggestions for implementing this within a mathematics class, for example [95] and [99]. In the remaining portion of the meeting, we look at instructor behaviors that encourage the students to direct their energies productively, and that foster the kinds of interactions that may lead to cognitive growth and meaningful understanding of the concepts of mathematics.

4.1   Description and Purpose of the Meeting

This meeting is intended to be a fifty minute session. The purpose of the meeting is to help instructors create successful in-class cooperative learning activities, by analyzing two issues critical to productive in-class group work. First, attention is paid to the design of activities that are conducive to group work. Instructors will analyze and comment on potential group activities. Second, the implementation of group work is analyzed via a case study. Typical problems and ways of circumventing them will be discussed.

4.2   Goals for the Meeting

1. To provide instructors with an understanding of the features of a mathematics problem that make it conducive to group work.

2. To alert instructors to difficulties that are commonplace in the implementation of in-class group work.

3. To give instructors specific strategies for dealing with the difficulties that can arise during the implementation of in-class group work.

4. To give instructors a forum for hearing the difficulties encountered and solutions tried by experienced instructors.

4.3   Preparation for the Meeting

1. Choose five or six representative problems from the textbook, most of which would be conducive to group work. Some features that make a problem conducive to group work include being too hard to do alone, being ill-structured, having more than one answer, requiring multiple ideas or viewpoints, and requiring interpretation or synthesis of concepts. Include at least one problem that you believe would not make a good group activity.

2. Ask those in your department who are experienced users of cooperative learning about their experiences with using groups in class. Try to determine the most typical problems with cooperative learning encountered by the instructors in your department. Gather ideas on successfully managing in-class groups. If necessary, revise the handout on "Managing In-Class Group Activities" to better suit the particular issues of your department.

3. Gather a panel of two to three experienced instructors who would be willing to share their experiences with the problems that can detract from group work.

4. Circulate the agenda for the meeting to all participants well in advance.

5. Make copies of meeting materials, and any questionnaires that you plan to distribute, and confirm attendance with participants.

4.4   Agenda for the Meeting (to be distributed to participants)

1. An analysis of textbook problems, and a discussion of the features that contribute to or detract from a problem being conducive to group work.

2. A discussion of in-class group management, and the keys to making groups work. The discussion will be triggered by a case study to be read.

3. A panel of experienced instructors will talk about their experiences with group management, and how they addressed the problematic issues.

4.5 Outline for Running the Meeting

:01–:05 Introduce the meeting, and briefly review the agenda.

:05–:20 On the chalkboard or overhead, list the following possible uses for in-class cooperative activities. Add any others that you see as appropriate.

- Shorter activities can be used to
  - check work on a routine problem,
  - clarify a concept through peer explanation, or
  - generate ideas when the class is stuck on a point.
- Longer activities can be used to
  - introduce concepts or skills,
  - clarify concepts or skills, or
  - practice and review concepts or skills.

This meeting will concentrate on the design and management of the longer cooperative activities.

It is easier, at least initially, to use (or modify for use) existing activities than it is to design new ones. In fact, most new textbooks have problems designed to be worked cooperatively. Some of the textbooks that we have personal experience with that include many exercises well suited for cooperative learning include the following.

- Calculus:
  - *"Calculus"* by Hughes-Hallett, Gleason, *et al.*
  - *"Calculus: Modeling and Applications"* by Smith and Moore
  - *"Calculus: Concepts and Contexts"* by Stewart
  - *"Calculus from Graphical, Numerical, and Symbolic Points of View"* by Ostebee and Zorn
  - *"Calculus, Concepts and Computers"* by Schwingendorf, Mathews, and Dubinsky
  - *"Workshop Calculus With Graphing Calculators : Guided Exploration With Review"* by Hastings and Reynolds
- Precalculus
  - *"Functions Modeling Change: A Preparation for Calculus"* by Connally *et al.*
  - *"Precalculus: Concepts in Context"* by Moran, Davis and Murphy
  - *"College Algebra from a Unified, Laboratory Perspective"* by Becerra, Sirisaengtaksin and Waller
  - *"Functions and Change: A Modeling Alternative to College Algebra"* by Crauder, Evans and Noell
  - *"Precalculus, Concepts and Computers"* by Dubinsky, Reynolds, et al

Pass out a handout with five or six problems from the textbook used in your department. Choose the problems so that most of them are good for use in a longer cooperative activity, but for different reasons. Also include at least one problem that you believe is not appropriate for an extended cooperative activity.

Ask the instructors to get into groups of three or four. Have them analyze the problems to decide which are appropriate for a cooperative activity. When most groups have reached a consensus, pass out the handout entitled "Good and Bad Features of Cooperative Activities." Facilitate a discussion on what features may imply that a problem would work well with cooperative learning, and what features should be avoided when choosing cooperative tasks. Use the list given in the handout as a starting point.

Emphasize that the main goal of cooperative learning activities is to encourage mathematical interaction. You may also note that fairly routine problems can be used in cooperative settings of smaller duration to accomplish the goals mentioned above. A more routine problem can also be modified, for example by removing some information or assumptions, or by generalizing the problem, to make it more conducive to an extended cooperative activity.

Finally, this is a good opportunity to discuss the importance of making intentional, goal-oriented decisions when planning a lesson. Have the instructors take one more look at each problem, and discuss what specifically they would want their students to learn from doing each problem. The process of using learning goals to plan lessons will be discussed in detail in Chapter 12.

The "Meeting Materials" section of this chapter includes examples of problems and reasons why these problems may be appropriate for use as in-class cooperative activities. This handout can be used by the meeting leader as a guide in planning this portion of the meeting, or it can be handed to the meeting participants during the meeting. More examples of problems suitable for cooperative learning are found in Chapter 4 of MAA Notes #37 [95].

:20–:35 Pass out the handout entitled "A Classroom Vignette," and allow participants a couple of minutes to read it. Lead a discussion based on the instructors' reactions to the vignette. Have them answer the following questions individually, and then lead a group discussion based upon their answers:

- What are your concerns about the way the activity was managed in the vignette?
- What could the instructor have done differently to make the management of the exercise more successful?

Some instructors may feel that they cannot make informed comments without seeing what actually occurred in the class. If some instructors feel that way, ask them to make their best determination on what the instructor could have done differently based on what was in the reading.

Pass out the handout "Managing In-Class Group Activities." Give the instructors a few minutes to read over the handout, and then discuss their reactions and observations.

:35–:45 Have the panel of experienced instructors talk about their experiences with group management, and how they addressed the problematic issues. If you do not assemble a panel of experienced instructors, you can alternatively allow the last ten minutes for the participants to get back into their groups to discuss and brainstorm solutions to problems they are facing in their own classrooms.

:45–:50 Wrap up the meeting with a word of thanks, and (as appropriate) time for participants' questions, a summary, suggestions for additional readings, or a questionnaire.

4.6   Meeting Materials

1. Copies of the handout entitled "Good and Bad Features of Cooperative Activities."

2. Copies of the handout entitled "Managing In-Class Group Activities."

3. Copies of the handout entitled "A Classroom Vignette."

4. Copies of five or six representative problems from the textbook that your department uses. (These will serve as examples for the instructor–trainees to analyze for appropriateness as cooperative activities.)

SAMPLE PROBLEMS FOR
IN–CLASS GROUP WORK

The following are problems that may be given to students in an introductory college mathematics class that uses in–class group work. Following each problem is a short discussion of the attributes that would encourage mathematical interaction among the students.

1. (Appeared as part of problem #11 on page 56 of Connally, *et al.* [31]) From December 1906 until July 1907, Bombay experienced a plague spread by rats. The table shows the total number of deaths at the end of each month.

Month	0	1	2	3	4	5	6	7
Deaths	4	68	300	1290	3851	7140	8690	8971

   (a) When was the number of deaths increasing? Decreasing?

   (b) For each successive month, construct a table showing the average rate of change in the number of deaths.

   (c) From the table you constructed, when will the graph of the number of deaths be concave up? Concave down?

   (d) When was the average rate of change of deaths the greatest? How is this related to the previous part? What does this mean in human terms as far as the spread of this epidemic is concerned?

   This problem requires the synthesis of different mathematical viewpoints (numerical, graphical, verbal), explanations of concepts, and interpretation of the mathematics in a context.

2. (Appeared as problem #2 on page 75 of Larson, Hostetler, and Edwards [73]) Describe the difference between a discontinuity that is removable and one that is non-removable. In your explanation, give examples of the following.

   (a) A function with a non-removable discontinuity at $x = 2$.

   (b) A function with a removable discontinuity at $x = -2$.

   (c) A function that has both of the characteristics described.

   This problem is ill-structured in that it doesn't make it clear if it is asking for a formula or merely a graph. There are many possible answers. Finally, it requires explanation of the concept.

3. (Appeared as problem #24 on page 262 of Hughes-Hallett, Gleason, *et al.* [61]) Let $f(v)$ be the amount of energy consumed by a flying bird, measured in joules per second (a joule is a unit of energy), as a function of its speed v (in meters/sec). See the figure below. Let $a(v)$ be the amount of energy consumed by the same bird, measured in joules *per meter*.

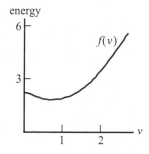

   (a) Suggest a reason for the shape of this graph (in terms of the way birds fly).

(b) What is the relationship between $f(v)$ and $a(v)$?

(c) Where is $a(v)$ a minimum?

(d) Should the bird try to minimize $f(v)$ or $a(v)$ when it is flying? Why?

This problem is quite difficult. It requires interpretation in light of a certain amount of knowledge about the "real world." Finally, it is an optimization problem, but it is not most easily solved by the methods typically employed for solving optimization problems in a calculus course.

4. (Appeared as problem #20 on page 137 of Osterbee and Zorn [86]) Sketch the graph of a continuous function g that has *all* of the following properties:

   (a) The domain of g is the interval $[-3, 5]$

   (b) $g''(x) < 0$ for all $x \in (-2, 1)$

   (c) $g''(x) > 0$ for all $x \in (1, 2)$

   (d) $g'(-2) < 0$

   (e) $g'(4) > 0$

   (f) $g'(1) = 0$ but $g(1)$ is neither the maximum nor the minimum value of g over the interval $[-2, 4]$

Because this problem has many parts, all of which must be considered together, it cannot be divided up among the group members. It helps to have many eyes and minds keeping track of all the features. Finally, there are many possible answers.

5. (Appeared as problem #15 on page 47 of Borelli and Coleman [17]) During a steady snowfall, a man starts clearing a sidewalk at noon, shoveling the snow at a constant rate and clearing a path of constant width. He shovels two blocks by 2pm, one block more by 4pm. When did the snow begin to fall? Explain your modeling process.

Additional assumptions are required to solve this problem. It involves mathematical modeling, and it also requires explanation of the modeling process.

Good and Bad Features of Cooperative Activities

- Features that may make a problem conducive to an extended cooperative learning activity include

  - too hard to do alone

  - ill-structured

  - more than one answer

  - useful to have multiple ideas or viewpoints

  - interpreting concepts

  -

  -

  -

- Features that may make a problem *not* conducive to an extended cooperative learning activity include:

  - too easy

  - mostly computational

  - calculator too much of a focus, becomes a barrier

  - easy to split into parts, or to "divvy up"

  -

  -

  -

Managing In-Class Group Activities[1]

The following is a list of some instructor behaviors that contribute significantly to the success of in-class cooperative learning activities.

1. The instructor gave clear directions on what to do during the group activities.

2. The instructor gave clear instructions on how and where groups should form, and what problem they should work on.

3. The instructor set a definite pace for the group work, and gave the groups a definite time frame in which to work.

4. As the groups worked on their problem, the instructor visited each group, and gave suggestions on what approaches they may try.

5. When wrapping up the group exercise, the instructor had the students explain the answers.

6. When concluding a group activity, the instructor briefly listed the vital points that were touched on in that activity.

Specific suggestions for implementing a cooperative activity include

1. **The mechanics of starting a group activity:**

   - Give very clear instructions on what activity you want the students to work on, and what you expect them to produce at the end of the activity. You can help get students focused by writing these instructions on the board, or putting them on an overhead. It is also useful to give them a brief account of the tasks that they will be undertaking in the exercise.

   - Immediately ask that one person in each group read the problem out loud to their group.

   - Occasionally ask all the groups to do the entire exercise at the blackboard.

   - When beginning the group activity, make a quick initial tour of the class to visit each group. If you see off-task behavior, or groups who are not getting underway, reiterate to them exactly what you want them to be doing.

2. **Dealing with off-task behavior:**

   - If you have persistent problems with groups working on their homework during class time, then break the class up into new groups before beginning an in-class group activity. You can break the class up by having them count off, grouping alphabetically, or grouping by birth date.

   - Try to be aware of when students are actually working on the activity you have assigned, and when they are doing something else. The following are some ways to encourage groups who are not working on the assigned activity:

     (a) Ask them which parts of the activity they have managed to work out, and have them explain one of their answers to you.

     (b) Tell them that you will be calling on their group to supply an answer when you wrap up the group activity.

     (c) Have their group move to the blackboard to continue their work. They will be more likely to make a serious attempt since their work will be visible to the entire class.

   - Plan additional activities for groups who finish the work very quickly. These do not have to be explicitly introduced or wrapped up. In many cases the people who regularly finish early will be very good at recognizing the value of activities for themselves. When you notice people finishing early, simply announce another problem to go on with, and write the reference on the board.

---

[1]Adapted from [38]

3. **Time management:**

   - Communicate, in advance, how long you have decided to spend on a group activity. Simply announce the allotted time before beginning the exercise. Giving the class a definite schedule can motivate them to get on with the activity more quickly.

   - If the group activity is a long one with multiple parts, break up the group work into sections. For example, when most groups have finished parts (a) and (b), stop the class and discuss only these parts. When finished, have the students go back to the rest of the problem. This can help slower groups keep up, and it can help maintain the desired pace in the group activities.

   - Another idea is to appoint a facilitator in each group who, in addition to participating in the group work, has the additional responsibility of making sure that the group sticks to the schedule.

4. **Facilitating the group process:**

   - Try to be pro-active when the students are assigned a group activity. You could establish ground rules governing what conduct is acceptable during in-class group activities. Take every opportunity to communicate these ground rules to the students, and make it clear that you think it is important for groups to function well in class.

   - If you have students move to different parts of the classroom to work, make sure that they sit so that group interaction is possible. Four people facing each other may encourage interaction; four people sitting in a row may not.

   - Once again, do not tolerate off-task behavior.

   - Do not allow individuals in the class to monopolize your time with individual concerns or issues that are irrelevant to the task you have assigned. Defer questions on grading and assessment to less public forums, or require that students put their questions on grading and assessment in writing.

   - Do not get too involved with any one group. Try to get to groups that need your assistance as quickly as you can. If you cannot get to them quickly, you can acknowledge that they are stuck, and promise that you'll be there as soon as you can.

   - Look for students who are obviously not working on the assigned activity, and encourage them to cooperate. If you see one person sitting out, ask that person to explain the group's answers to you. Encourage people in groups to compare answers with and to explain their methods for doing the problem to the other people in their group.

   - Do not be too directive in your assistance with the mathematics. Encourage the group process by facilitating dialogue among the members. Please note that we are not recommending that you never answer the students' questions. We are, instead, suggesting that you first determine if another student is able to answer.

A Classroom Vignette[2]

The following is a brief account of one instructor's use of a cooperative learning opportunity in an elementary mathematics class.

The instructor started class by spending twenty minutes returning papers and making administrative announcements. The instructor then began the lesson by reviewing the topic treated in the previous class: the relationship between concavity and the rate of change of the slope. The instructor lectured for ten minutes, and occasionally posed questions to the class.

To follow up on the lecture, the instructor assigned a problem from the text, and told the class to break into groups and begin working. Most of the students were sitting with the members of their homework teams, and so these were the people to whom they turned. Many of these groups engaged in spontaneous and long conversations, discussing the problems from the homework assignment due later in the week. In the meantime, the instructor walked around from group to group. Despite the presence of the instructor, many groups remained focused on their homework, rather than on the assigned activity.

As the activity went on, some groups finished their off-task conversations, and looked around for some clues as to exactly what they should be doing. One student got up and walked out of class. As the instructor passed a group, one student asked a question about the material to be covered on the quiz. The other people in the group were interested in this, and stopped their activity to listen. The instructor smiled and tried to answer the question without revealing which questions would be on the quiz.

While the instructor chatted, other students had raised their hands. Some groups finished working on the assigned activity, and looked around for something to go on with. Finding nothing, they talked about what was going on in their lives. One student eventually became tired of waiting, and began a very loud conversation on the subject of the last weekend's football game.

The instructor then walked to the front of the room, and called the class to order. Standing at the front of the room, the instructor had some of the groups read their answers to the class. When they were finished, the instructor began to lecture.

---

[2] Adapted from [38]

4.7 Suggested Reading

- Matthew DeLong and Dale Winter. "Addressing Difficulties with Student-Centered Instruction." [38]

  Here we identify many problems that plague some in-class groups, and we offer many practical suggestions for dealing with these problems. The main problems that we address are inattention to the instructor and other students, off-task activities, and transitions to and from group activities.

- Barbara Reynolds, Nancy Hagelgans, Keith Schwingendorf, Draga Vidakovic, Ed Dubinsky, Mazen Shahin, and G. Joseph Wimbish, eds. *"A Practical Guide to Cooperative Learning in Collegiate Mathematics. MAA Notes #37."* [95]

  This book presents a theory of cooperative learning that revolves around groups that stay together for the course of an entire semester, and the discussion is relevant for either in-class groups or after-class groups.

  In Chapter 6, the authors describe a number of areas that students struggle with when they try to work in cooperative groups. The authors classify these as difficulties connected with (1) modes of operation, (2) organizational issues, (3) group dynamics, (4) individuals and (5) the rest of the world. As the breadth of some of these categories may suggest, some of the difficulties are simply too general to offer useful, concrete advice, and others may stem from deeper cultural issues that are very difficult to address in any event, and perhaps particularly so in the context of math class. The authors' suggestions for dealing with difficulties primarily revolve around prevention by trying to convince the students of the efficacy of cooperative learning, and intervention by regular meetings between the instructor and the groups.

  Chapter 4 describes what cooperative learning groups do as they learn mathematics. The kinds of problems are broken down into *activities*, which are done as an introduction to a concept, *tasks*, which build upon and synthesize ideas developed in activities, and *exercises*, which reinforce the concept being studied. Examples are given of each type of problem, and the examples are drawn from courses throughout the mathematics curriculum.

- Neil Davidson. "Small Group Learning and Teaching in Mathematics: A Selective Review of the Research" in MAA Notes #44. [43]

  Davidson reviews those articles he sees as relevant to mathematics teaching. There is a fair amount of useful information here. For example, a list of the typical ingredients or phases for a cooperative learning activity, guidelines for small-group examinations, etc. There is also a reasonable summary of the wide variety of conceptions of exactly what "cooperative learning" is, and how different people have attempted to implement it. There are some reports of student learning outcomes from cooperative learning, although the reports are very brief, and it is not usually clear under what circumstances the results were collected.

- David Johnson and Roger Johnson. "Social Skills for Successful Group Work" in MAA Notes#44. [43]

  This article appears to be aimed at elementary school teachers. The point that it makes is that students may not instinctively know how to function "well" in a cooperative learning situation, and that appropriate behaviors need to be recognized and acknowledged by the teacher. Johnson and Johnson suggest some devices for doing this, although many of the specific devices suggested are not suitable for college classrooms.

- Patricia Heller and Mark Hollabaugh. "Teaching Problem Solving Through Cooperative Grouping Part 2: Designing Problems and Structuring Groups" in MAA Notes #44. [43]

  This article describes the use of in-class groups in elementary college physics courses. In the second part of the article, Heller and Hollabaugh discuss three questions concerning the formation and structure of cooperative groups. These are: (1) What is the "optimal" group size for physics problem solving? (2) What ability and gender composition groups result in the best problem-solving performance? and (3) How can problems of dominance by one student and conflict avoidance within a group be addressed? The article describes two approaches: (1) assigning roles to group members, and (2) discussing group functioning.

- Elizabeth C. Rogers, Barbara E. Reynolds, Neil A. Davidson, and Anthony D. Thomas eds. *"Cooperative Learning in Undergraduate Mathematics: Issues, and Strategies that Work.* MAA Notes #55."* [99]

  This volume of the MAA Notes series describes a variety of specific techniques for implementing in-class group work, written specifically for the college mathematics instructor. Unlike many of the articles that have appeared on the use of cooperative learning in mathematics over the last ten years, this book describes some of the theories of how students can learn in a group setting, although the focus is on using theories of learning to design activities that promote positive student-student interactions, rather than on helping instructors decide what to do as they attempt to manage an interactive classroom. The book briefly discusses some of the difficulties that may be encountered when using in-class group work. Most of the advice offered has to do with attempting to prevent such difficulties from the outset by deliberately trying to foster a positive climate within the classroom. This book includes a fairly extensive bibliography including textbooks and other instructional materials that are very highly compatible with the use of cooperative learning in college mathematics classes.

Chapter 5

Getting Students to Read the Textbook

Observing textbooks on introductory collegiate mathematics, and instructors who use these textbooks, one notices that there is little consensus on the role or function of a mathematics textbook. Some textbooks attempt to provide a narrative that tries to access concepts through intuition, application, and explanation. Others attempt to provide an exhaustive resource and reference. Yet others function primarily as a repository of worked examples. In addition, there are the ubiquitous tensions between rigor and accessibility, and between conceptual understanding and skill acquisition.

In our experience, many students use mathematics textbooks as little more than large collections of homework problems, templates for the "types" of problem encountered in the homework, and lists of answers (usually to the odd-numbered exercises). In some of the classes that we have taught, students clearly believed that their textbooks were dry, boring, expensive, and not at all to be read!

We both base our philosophies of the role of the textbook in a mathematics course on the assumption that such an expensive investment ought to, in fact, be read. Much more than economic considerations drive our intent to have our students read their textbooks—we feel that the textbook can and should be a vital part of the course that contributes to the learning goals that we have set for our students. Many of these specific goals are mentioned in this meeting.

We base our own teaching on the assumption that students have already read the section under consideration *before* coming to the day's class. We do not expect a complete understanding of the material, and we acknowledge that even if all students read the book they will come to class with different levels of apprehension of the material. However, we believe that all students can, and should, come to class having encountered the basic vocabulary and techniques, having thought a bit about the concepts to be studied, and having reflected on their own initial understandings and where those need to be clarified or solidified. We then run our class sessions as if the students have come prepared, which allows us to more quickly delve into substantial, challenging, and interesting aspects of the material. We then assign homework problems to be done after the class lesson, and view these problems as the students' opportunities to assess their own understanding of the reading and class session. In other words, our model resembles:

$$\text{read} \rightarrow \text{class} \rightarrow \text{homework}.$$

Because of the lack of general agreement on the role of a mathematics textbook, we begin this meeting with considerations of why students should read the book, and this is intimately related to a discussion of course goals and philosophies. Because most college students do not know how to read a *mathematics* book, we follow with an activity that is designed to help instructors convey to their students this vital skill. Finally, because any course component that is stated to be of importance should be often assessed and occasionally evaluated, we close the meeting with ideas about how to assess reading.

5.1 Description and Purpose of the Meeting

This meeting is intended to be a fifty minute session. The purposes of the meeting are to emphasize the importance of reading the textbook within the larger objectives of the course, and to help instructors find ways to encourage their students to read the textbook regularly and skillfully. The meeting consists of discussions of why and how the students should read a mathematics textbook, as well as strategies for teaching students how to read the textbook. Strategies for encouraging the reading are considered. Finally, techniques for assessment of student reading are discussed.

5.2 Goals for the Meeting

1. To help instructors identify the function of reading the textbook within the framework of the course.

2. To help instructors understand their own behaviors that encourage or discourage the students from reading the textbook.

3. To give instructors the opportunity to clarify how to read a mathematics textbook, and how to teach their students to do so.

4. To give instructors the opportunity to discuss techniques for assessing their students' understanding of their reading.

5.3 Preparation for the Meeting

1. Select two short sections out of your course textbook that may be difficult for students to read and make copies for the meeting participants.

2. Circulate the agenda for the meeting to all participants well in advance.

3. Make copies of meeting materials, and any questionnaires that you plan to distribute, and confirm attendance with participants.

5.4 Agenda for the Meeting (to be distributed to participants)

1. Activity and discussion on how reading the textbook fits into the larger goals of the course.

2. Discussion of instructor behaviors that encourage or discourage reading the textbook.

3. Activity designed to help instructors teach students to read a mathematics textbook.

4. Brainstorming session on assessment techniques that will encourage students to regularly read the textbook.

5.5 Outline for Running the Meeting

:01–:05 Introduce the meeting, and briefly review the agenda.

:05–:15 Pass out the handout entitled, "What Do I Mean by 'Read the Book'?" and ask the instructors to fill in their responses. Based on their ideas from the handout, have the instructors discuss exactly what they want the students to get from reading the book. You may have them think about what an ideal book would contain, and how it would be structured. Write a list on the board of reasons why it is important for students to read the textbook. Some possibilities are that

- some studies show reading is better than lectures for retention [81];
- using multiple approaches to teaching the material can accommodate different learning styles;

- there is a limited amount of class time to "cover material" in appropriate depth;

- introduction to basic vocabulary and skills can easily be acquired by reading the textbook, thereby leaving class time for clarification, extension, and reinforcement of the material;

- a textbook may take a more narrative and intuitive approach to introducing concepts, which may appeal to introductory mathematics students;

- access to more than one narrative on the material—the instructor's and the author's—can provide for a broader and more robust understanding of the material;

- the graphics and figures interspersed within the text facilitate connection among multiple representations—graphical, numerical, symbolic, and verbal; and

- reading technical material is a valuable transferable skill for future employment and for life-long learning.

Emphasize that it is important that the instructors communicate to their students very early in the semester why the students should read the book, and what the instructor expects when he or she says, "Read the book." The instructors could use this opportunity to re-articulate the goals of the course, and point out to the students the ways in which reading fits into these goals.

At this time, communicate to the instructors the role of the textbook within the framework of the specific course goals at your institution, and how reading relates to the other components of the course—such as lectures, activities, homework, and exams. Some sample course goals and the way the textbook may support them include the following.

- To enable students to write coherently about mathematics and to give clear verbal explanations of their mathematical thinking:
  - the textbook may support this goal by giving students well-written examples of explanations of concepts aimed at their level of understanding.

- To improve students' attitudes towards mathematics:
  - the textbook may support this goal by making mathematics interesting and accessible, and by giving examples of applications of mathematics to real world situations.

- To improve students' abilities to translate word problems into mathematical statements and back again:
  - the textbook may support this goal by providing thoroughly explained examples of mathematical modeling of real world phenomena.

- To improve students' abilities to translate among mathematical representations (numerical, graphical, symbolic, and verbal):
  - the textbook may support this goal by providing multiple representations of problems and by emphasizing the connections among representations.

- To strengthen students' sense of responsibility for their own learning:
  - the textbook may support this goal by providing the students with an opportunity to learn the material independently.

- To encourage students to see mathematics as a way of thinking and not merely as a collection of rules and formulas:
  - the textbook may support this goal by showing mathematics as a process that involves exploration, conjecture, testing, and proof.

:15–:20 Give a brief overview of the kinds of classroom activity that encourage students to read the textbook (assessment techniques will be discussed later in the meeting). Remind the instructors of the contents of the handout "Getting Students to Read the Book" from the pre-semester orientation session, Block B (cf. Chapter 3). If you did not use the handout in the pre-semester training, you may pass it out during this meeting. Relate the handout to the previous discussion. Make sure that the instructors know that they must continue to communicate, verbally *and through their behavior*, that they expect the students to read the textbook. Moreover, continue to emphasize that the instructors should not lecture on all of the material as if the students have never before seen it. When structuring their class activities, they should assume that the students have done the assigned reading!

As an example of that last point, consider the different ways a lesson on linear functions may look. If you do not assume that the students have read the textbook, then you likely would start with a definition of a linear function, and the derivation of several equations of lines. On the other hand, if your students have read the book, you could assume that they know the basic form of a line, and how to find a slope and intercept. You could solidify these ideas while developing deeper understanding by starting with a modeling problem that requires finding the formulas and intersection point of two lines from a word problem.

As another example, consider a lesson on polar coordinates. If you assume that your students have not read the section, you would likely begin by defining polar coordinates and putting the conversion formulas on the board. If you assume that your students have a basic familiarity with the idea of polar coordinates, you could immediately precede to asking them to convert some points from polar to rectangular coordinates, or to asking them to plot the graphs of simple equations like $r = 2$ or $\theta = \frac{\pi}{4}$.

A possibly more radical approach that you could bring up with your instructors has been advocated by some. Rather than merely designing activities and short lectures that assume the reading, some instructors devote their entire class time [3] or a substantial amount of class time [54] to student questions and questioning students on the reading. Beginning instructors may be reluctant to move very far in this direction, but the methodology could be easily incorporated on a small scale.

:20–:35 Lead an activity on how to help the students learn to read a mathematics textbook. This activity is designed to identify specific behaviors that are helpful when reading a mathematics textbook, and to demonstrate one way of teaching the students these behaviors.

Begin by leading a five-minute discussion on how the instructors read a mathematics text in their own fields of research or interest. Make sure to contrast this with how they read other forms of print, such as novels or newspapers. Culminate the discussion by making a list on the board of specific behaviors that students should engage in when reading their mathematics texts. Here it would be most appropriate to make the traditional observation that the most useful way to read mathematics is with pencil and scrap paper at hand. Some behaviors useful when reading mathematics are

- making marginal notes;
- justifying successive steps of a calculation or proof;
- identifying assumptions or hypotheses;
- working out examples *before* looking at their solutions in the text;
- making a separate list of unanswered questions;
- *re*-reading until understanding is gained;
- making summaries or outlines identifying main ideas;
- relating the current section to previously encountered material;
- making a Content, Form, Function outline [4];
- continually asking what the purpose of the current discussion is; or
- reading non-linearly.

Once you have identified these behaviors, divide the instructors into pairs. Give each pair one copy of both of the sections of the book that you selected. Ask each instructor to take a few minutes to read one of the sections by specifically engaging in the listed behaviors they see as applicable. When they have finished reading, ask them to alternate role-plays of an instructor and a class. Ask the person role-playing the instructor to read the section aloud to the person role-playing the class, while appropriately pausing and explaining when and how to engage in the listed behaviors. This may seem to be quite artificial and odd to the instructors while they are role-playing. It may be helpful to remind the instructors of their goal: to develop a clear and purposeful demonstration of the techniques of engaged mathematical reading, suitable for undergraduates. The role-play should take about five minutes, and then the instructors should switch their roles and repeat the activity.

Because the material of the section of the textbook will not be a challenge to the instructors, they may need to be reminded that their own comprehension of the material will differ vastly from that of their typical students. Therefore, as is the case most of the time when teaching, the instructors must view this activity through the lens of their students' relationships with the material and not their own.

Some instructors may be quite resistant to the idea of participating in a role-play. Listed below are some strategies for getting reluctant instructors to participate.

- Acknowledge that the situation is artificial; nevertheless, it offers an opportunity to practice skills and receive feedback on them.

- Point out that role-playing in this way is an opportunity, not an obligation, but it is an opportunity that can only work if everyone participates.

- If someone has a serious problem being either the student or the teacher, assign them to be an observer of a pair. The observers can take notes as to what things the "instructors" did when teaching their "students" how to read the textbook.

If you think that your group of instructors will be unwilling to participate in a role-play, you could have an experienced instructor role-play the activity for the entire group. However, this has the disadvantage that none of the participants will have the opportunity to practice for themselves teaching students to read the book.

Wrap up this segment of the meeting by encouraging the instructors to actually engage in this reading–modeling activity in their own classes at least once early in the semester.

:35–:45 Lead a discussion designed to generate ideas among the instructors for assessment techniques that can be used to encourage reading of the textbook. Make sure to emphasize that to encourage higher-level reading skills, assessment must be of higher-level thinking, and not just factual retention. See Bloom's taxonomy [12] for a classification of levels of thinking skills and levels of questions.

You could start the discussion by passing out the handout entitled "Reading Assessment Techniques" in case the instructors do not have any fresh ideas at this point. This handout provides a list of some assessment ideas.

In your discussion you can point out the various roles that such assessment could play, beyond merely being a motivational tool, or a tool for assessment. For example, you could emphasize using assessment techniques that are themselves learning experiences. For example, requiring the students to submit notes taken on a section may be preferable to having a short quiz on the reading. As mentioned previously in the meeting, when the assessment techniques are also learning activities, these can be vitally used during class time as another substitute for lecture.

:45–:50 Wrap up the meeting with a word of thanks, and (as appropriate) time for participants' questions, a summary, suggestions for additional readings, or a questionnaire.

5.6 Meeting Materials

1. Photocopies of two short sections of the departmental textbook.

2. Copies of the handout entitled "What Do I Mean by 'Read the Book?' "

3. (If you did not already use it during pre-semester training) Copies of the handout entitled "Getting Students to Read the Book." See the "Meeting Materials section of Chapter 3.

4. Copies of the handout entitled "Reading Assessment Techniques."

What Do I Mean by "Read the Book?"

Complete the following statement by circling as many phrases as you like.

When I say, "Read this material before our next class," I mean that students should:

memorize definitions and theorems   understand proofs

think about concepts

understand examples   have questions

skim the material   be ready to do applications

be ready to discuss concepts   be ready for a quiz

know why the material is important

know how the material relates to previous reading

be able to explain the material   be able to do routine problems

be able to summarize the material

be able to use new terms   be able to do complex problems

What else?

This exercise was adapted from one developed by Dianne Holt-Reynolds, Michigan State University.

Reading Assessment Techniques

- Have the students write daily reading journals that are periodically read and commented upon by the instructor [54].

- Have the students submit daily questions on the reading—both those that were answered for them in the text and those that were left unanswered [54], and use the submitted questions to hold a question and answer session [3].

- Have the students submit section summaries, either individually or in pairs.

- Have the students E-mail difficulties and questions prior to class, and then use these to drive the classroom activities [90].

- Pre-assign two or three questions for the students to answer in their reading, and then give periodic quizzes on these questions.

- Begin class with a brief vocabulary quiz, in which the students are asked to explain a new term or concept in their own words.

- Require the students to take notes on the reading, and collect and assess these notes [78].

- Assign homework on material that was not covered in class [33].

- Test your students' abilities to read mathematics by putting some mathematics to read on a test, and then asking them a question that (in theory) only those who read and understand the section can answer. In order to test their reading abilities and not their prior knowledge, the mathematics should be accessible yet new to them [33].

5.7 Suggested Reading

- The MAA's Innovative Teaching Exchange has a collection of articles available from: `www.maa.org/ t_and_l/exchange/exchange.html`.

  The articles included here that pertain to reading are: "Math Class—Have You Seen the Preview?" by H. Louise Amick [2], "Requiring Student Questions on the Text" by Bonnie Gold [54], "How I (Finally) Got My Calculus I Students to Read the Text" by Tommy Ratliff [90], "Helping Undergraduates Learn to Read Mathematics" by Ashley Reiter [94], and "Volumes and History: A Calculus Project Involving Reading an Original Source" by Elyn Rykken and Jody Sorensen [100].

  Amick describes an approach for changing a class from a lecture class into one focusing on problem-solving and discussion through reading previews. The students are asked to read a preview before each lesson, in order that every student comes to class ready to contribute to the development of the lesson. These previews focus on recall of prerequisite knowledge as well as discovery of upcoming content through graphing calculators or well-chosen examples.

  Gold describes how she requires student questions on the reading. She also describes how she assesses and uses the questions. Finally, she gives one approach to helping students learn to read mathematics—the Content, Form, Function outline advocated by Angelo and Cross [4].

  Ratliff gives examples of readings that he assigns. He discusses the advantages and disadvantages of assessing these via in-class quizzes or E-mail. He points out one advantage to having students E-mail mathematics, that is the lack of symbols forces students to find their own words to explain what is going on rather than resorting to answers in the form of formulas.

  Rieter gives some advice for students on how to read theorems and how to read definitions.

  Rykken and Sorensen give an example assignment that they have used that requires students to read an original source. They point out the pedagogical advantages, which include increasing students' knowledge of mathematical history as well as changing students' conceptions of the nature of mathematics.

- Carl C. Cowen. "Teaching and Testing Mathematics Reading." [33]

  Cowen gives advice for helping students learn to read mathematics by guided discussion in class. He suggests assigning sections that may otherwise be skipped to be read by students, followed by homework assignments on those sections. Finally, Cowen discusses how to test students' ability to read mathematics. The context is primarily one of intermediate courses, such as linear algebra.

- Underwood Dudley ed. "*Resources for Calculus Collection, volume 5. Readings for Calculus.* MAA Notes #31." [44]

  This volume provides supplemental readings on the history of calculus and of mathematics, on the nature of mathematics and its applications, on the learning of calculus, and on the place of calculus and mathematics in society. It could be used to provide reading material in a calculus course that uses a traditional, less narrative-oriented, textbook.

- Articles by Paula Maida [78], Janet Andersen [3], and Mark Johnson [63] in "Instructor's Resource Manual: Calculus from Graphical, Numerical, and Symbolic Points of View" by Osterbee and Zorn.

  Maida gives a rationale for requiring students to read the textbook before the lesson on that particular section. She explains how she requires students to take notes on pre-read sections. She also includes some student reactions to this teaching technique.

  Andersen explains how she structures her entire classroom methodology around student questions and questioning students. A substantial aspect of this involves the students reading the material before coming to class.

  Johnson writes a short note to the students giving his advice on how to read mathematics textbooks.

- Matt Boelkins and Tommy Ratliff. "How We Get Our Students to read the Text Before Class." [14]

  The comments on this article for the suggested reading are: In this article, the authors point out that when students read the text before class, the fundamental nature of the class is transformed from the introduction and illustration of very basic concepts to working with the major ideas, interesting applications and really difficult mathematical issues. The authors also point out that students may gain a greater sense of independence, and that this can make for livelier student-student interactions during class. The authors note some of the problems: students generally have little experience with reading their texts for understanding, students' patterns of work utilize the text more as a reference rather than a primary source, and that many students are not able to consistently complete reading assignments without some powerful form of extrinsic motivation, i.e., grades. The authors describe their approach to encouraging students to read the text. This consists of setting reading goals and making a set of questions that students complete when they have finished the reading and then e-mail their responses to the instructor before the next class period. The authors also indicate some of the ways in which they have been able to tailor their instruction based on the students' responses, and describe some student reactions.

Chapter 6

Assessing and Evaluating Students' Work

The mathematics course that we envision utilizes a variety of assessment techniques, including team homework, projects, student presentations and writing assignments. A common element of these assessment practices is the clear and precise communication of mathematical ideas.

In many cases, students are expected to work cooperatively – to produce a single piece of work representing the collective efforts of three or four students. In order to provide sufficient challenge for a group of students and sufficient incentive for the students to work cooperatively (instead of simply dividing the work among themselves, and later compiling their individual contributions for submission) the problems assigned are more complicated, and are sometimes more "open-ended" than exercises typically assigned in traditional mathematics classes. The problems assigned typically require students to make appropriate assumptions, to try alternative avenues of inquiry, to try to understand the mathematics more thoroughly by recognizing its application to otherwise unfamiliar situations, etc.

The work that students produce on these more complicated assignments is not simply pages of algebraic manipulations with boxed answers at the end. Instead, students are encouraged (and helped) to exhibit their understanding of the mathematics in multiple ways (such as graphs, written accounts of their assumptions and reasoning processes), instead of simply recording the algebraic steps that they performed.

As may be expected, assessing and grading student work of this kind can be radically different from assessing and grading pages of algebraic manipulations with a conveniently highlighted answer at the end. Questions that instructors may ask themselves range from, "Should I deduct points for poor spelling or grammar, even though the students got the right answer?" to "This is so much more work. Why should I bother?" In this meeting we hope to create an experience, and a forum for discussion, that can help instructors develop the skills and judgment that they will need for assessing these more complicated homework assignments.

There may be some overlap between the content of this meeting and the pre-semester training provided to instructors. In particular, if you have used the pre-semester orientation sessions described in Chapter 3, then instructors will already have had some introduction to the issues involved in grading students' work. With this in mind, we have tried to provide materials that are highly adaptable. The meeting materials included in this chapter include samples of student work (to practice grading) in the form of short, written responses (suitable for instructors with little or no experience), and longer report-style student work (suitable for more experienced instructors who are already comfortable with grading less complicated examples of student work).

6.1 Description and Purpose of the Meeting

This meeting is intended to be a fifty minute session. The main purpose of this meeting is to provide an authentic experience of grading student work, and to suggest some ideas for how instructors can go about grading in a systematic and consistent fashion. A secondary purpose is to help instructors to realize that the feedback that they provide to students is important. This experience is intended to be thought-provoking. The participants should be encouraged to think about how their grading methodology contributes to their goals for their class.

6.2 Goals for the Meeting

1. To provide instructors with an opportunity to practice grading samples of student work before needing to do this for their own classes.

2. To provide instructors with an opportunity to think about the criteria that they will use to assign points for student work.

3. To provide instructors with an opportunity to deal with different student interpretations of the meaning and requirements of problems, and an opportunity to think about how they will deal with these kinds of situations.

4. To provide instructors with an opportunity to think about what is important to them in students' solutions.

5. To provide instructors with some ideas on how they can go about assessing and grading non-traditional assignments in a systematic way.

6. To provide instructors with some advice for providing feedback to students.

7. To provide instructors with a forum for hearing others' thoughts on the issues outlined above.

6.3 Preparation for the Meeting

1. Gather institutional or departmental guidelines on grading, if any such guidelines exist.

2. Find out what kinds of student work instructors will have to grade, and what kinds of assignments that instructors feel ready to grade.

3. Select a problem to use as the focus of the grading session. If instructors are less experienced and less confident in their abilities, select a simpler problem; if instructors are more experienced and confident, select a more involved problem.

4. Collect samples of student solutions for this problem. Be sure to get the students' permission to use their work for the meeting. If student work is not available, create several different sets of solutions, or else use the samples included in the "Meeting Materials" section of this chapter. Be sure that the solutions are not all perfect.

5. Circulate the agenda for the meeting to all participants well in advance.

6. Make copies of meeting materials, and any questionnaires that you plan to distribute, and confirm attendance with participants.

6.4 Agenda for the Meeting (to be distributed to participants)

1. Outline of departmental and institutional regulations governing testing, grading and homework.

2. Practice grading session on samples of student work.

3. Review and discussion of grading guidelines and suggestions for providing students with feedback on their work.

6.5 Outline for Running the Meeting

:01–:05 Introduce the meeting, and briefly review the agenda.

:05 to :25 Distribute the copies of the problem and student solutions to instructors. Tell the participants what you want them to do, and how much time they have. This may include assigning points for each of the solutions, writing comments or making corrections to each of the solutions. The participants can grade individually, or, if you feel that it is helpful for them to discuss what they are doing, you may like to have them work in pairs. When the participants have finished grading, it is helpful to point out how much they have been able to achieve during the time allotted. The meeting leader may extrapolate the amount of time required for grading homework each week from this, and point this out to the participants.

:25–:30 Have participants report the results of their work. This may include reporting the scores that they have assigned to each of the pieces of work, and the reasons that they have assigned those scores. While participants report the scores they have assigned, the meeting leader can keep a tally of the scores on a chalkboard. Often, there will be one or two participants who have assigned very high or very low scores. These participants should be encouraged to share their reasons. It may also be useful for participants to describe the kinds of comments that they think would be appropriate to write on the students' work, and to have them say what they hope those comments would achieve. If themes emerge from the reports, the meeting leader can summarize these themes on a chalkboard for reference.

:30–:40 Distribute copies of the handout, "A Few Suggestions for Student Feedback." Give the participants a few minutes to read over them. If they have not done so already, ask the participants to write comments on the samples of student work that they have graded. Have the participants discuss their comments with a neighbor, paying particular attention to: (1) the messages (implicit and explicit) that the comments would likely convey to students, and (2) how the comments could help students to learn mathematics. When instructors have had a chance to talk this over, ask several of them to share what they have just learned about writing comments with the other participants.

:40–:45 Approach 1 If you are leading a meeting that is aimed at the needs of less experienced instructors and short pieces of student work, distribute the handout "Grading Rubrics" to the participants. There probably will not be enough time for a thorough discussion of both the generic and specific approaches. Select one part of the student work that the participants graded earlier, and ask the participants to use the procedure illustrated in Figure 1 to create a grading rubric for that particular part of the problem.

:40–:45 Approach 2 If you are leading a meeting that is aimed at more experienced instructors and longer, report-style student work, distribute the handout, "Assessment Issues for Student Writing." Once instructors have had a few minutes to read over the issues, ask very experienced instructors to comment on the issues. In particular, ask for comments on either issues not listed on the handout, or ways of addressing these issues that the experienced instructors have tried.

:45–:50 Wrap up the meeting with a word of thanks, and (as appropriate) time for participants' questions, a summary, suggestions for additional readings, or a questionnaire.

6.6 Meeting Materials

1. Copies of institutional or departmental guidelines or regulations covering grades.

2. Copies of a typical homework problem.

3. Copies of two or three different student solutions to the homework problem.

4. Copies of written grading guidelines: "A Few Suggestions for Student Feedback," and "Grading Rubrics."

5. If you plan to focus your meeting on the assessment of student writing, copies of the handout, "Assessment Issues for Student Writing."

Example Problem: Students are expected to provide some short, written responses [1]

In month $t = 0$, a small group of rabbits escapes from a ship onto an island where there are no rabbits. A study determines that $p(t)$, the island rabbit population in month t, is given by

$$p(t) = \frac{1000}{1 + 19(0.9)^t}, t \geq 0,$$

(a) Evaluate the expressions $p(0)$, $p(10)$, $p(50)$, and explain their meaning in the context of rabbits.

(b) Sketch a graph of $p(t)$, $0 \leq t \leq 100$, and describe the graph in words. Does it suggest the kind of growth you would expect from among rabbits on an island?

(c) Find $p^{-1}(50)$ and explain its meaning in the context of rabbits.

(d) Estimate the range of $p(t)$. What does this tell you about the rabbit population?

(e) Explain how to find the range of p algebraically.

---

[1] From *Functions Modeling Change.* by Conally, Hughes-Hallet, Gleason, et. al. Preliminary edition (1996)[31].

①

Team 6

Homework #2

Section 1.4 #23

(a) $p(t) = \dfrac{1000}{1 + 19(0.9)^t}$

$p(0) = 50$. 50 rabbits when t=0.

$p(10) = 130$. 130 rabbits when t=10.

$p(50) = 910$. 910 rabbits when t=50.

(b)

Since it is an island, there would be no natural rabbit predators on the island to keep rabbit numbers down (predators would only be there if there were rabbits to start off with, but there weren't any rabbits since it is an island). So, the rabbits

②

can just reproduce without being killed off. This is why the graph shows the shape of exponential growth.

(c) $t \xrightarrow{\ p\ } $ rabbits $\xleftarrow{\ p^{-1}\ }$

$p^{-1}(50) = t=0$

$p^{-1}(50)$ is the time when there are 50 rabbits.

(d) Since the graph showed exponential growth, the range is all y-values above 50.

(e) Procedure: ① Since $p(t)$ is a fraction function, find the x-values that make the bottom line zero.

② Take all x-values except the ones found in ①.

③ Figure out which y-values you can get from the x-values.

- 1 -

105 Group 1 Homework.

Problem 23.

(a) $p(0) = 50$. This means that at time 0 (when the rabbits are released) there are 50 rabbits.

$p(10) = 130$. This means that in month 10 after release, there are 130 rabbits (rounded off.).

$p(50) = 911$. This means that in month 50 after release, there are 911 rabbits (rounded off.).

(b) We got this graph on the calculator:

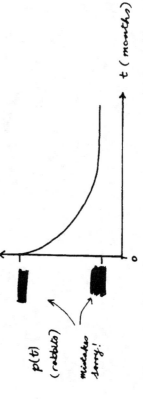

p(t) (rabbits)

mistake here!

t (months)

This graph shows that there are too many rabbits to start off with. Since this is an island, the rabbits

- 2 -

can't go anywhere else to find the things they need to survive (food, etc. so they die off until there are just the right number of rabbits for the island. This can be seen as the graph starting off high and then decreasing to a level number of rabbits.

1(c) As we discussed in class,

$$\text{months } t \xrightarrow{\ p\ } p(t) \text{ rabbits}$$
$$\xleftarrow{\ p^{-1}\ }$$

So, in words, $p^{-1}(50)$ is the month when there are 50 rabbits. From part (a), this is zero months.

(A) When we traced the graph, we got:
Range: $1000 < y < 1019$

k) How to find range algebraically (only a theoretical procedure):

(a) Find the x-values that can go into the formula (domain.)

(b) Based on the domain, find all the y-values that can come out of the formula (range).

Math 105 Group Homework - Group 3

(a) We evaluated the expressions as follows :

p(0)=50   This means that zero months after the rabbits escape, there were 50 rabbits.

p(10)=131.14   This means that 10 months after the rabbits escaped, there were about 131 rabbits (rounded since 0.14 rabbits doesn't make sense)

p(50)=910.81   This means that 50 months after the rabbits escaped, there were about 911 rabbits (rounded).

(b) Graph of p(t)

t	0	10	20	30	40	50	60	70	80	90	100
p(t)	50	131	302	554	781	911	967	988	996	999	999

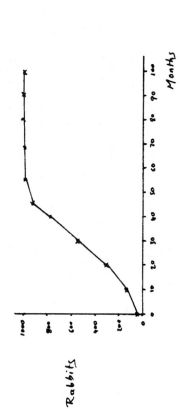

Growth of Rabbits on an island

Yes, because to start off with there are only 50 rabbits on the island. This is only 25 breeding pairs. We are not told how big the island is, but 25 pairs are probably not enough to over-populate the island too quickly. So, because there are only 25 pairs to start off with, the population would increase as these rabbits had babies, but it wouldn't increase too much because there are only 25 pairs. This is why the graph is going up, but not too fast, at the beginning. Next, when the babies grow up, they will start having families of their own, so there will be even more rabbits being born. This is why the graph starts to go up faster. However, as there are more and more rabbits, any creatures on the island who eat rabbits will increase as well, and more rabbits will be killed. Also, the grass and plants that the rabbits eat will be used up faster (since there are more rabbits eating them), so some rabbits may die of starvation. This is why the graph levels off.

(c) p⁻¹(50) = 0.   This is the time when there are 50 rabbits, i.e. t=0.

(d) When we graphed the function on the calculator, it didn't turn out properly, and we didn't find the right viewing window to get the same graph as part (b). The range that we got from the calculator was that the possible y-values were between 1000 and 1019. We couldn't trace our graph from (b) like on the calculator, but we figured that the y-values were between 50 and 999 (approx.)

This tells you that there is always between 50 and 999 rabbits on the island.

(e) To find the range algebraically, you find out the domain and then put this into the function to get the possible y-values. We couldn't find the domain, though. The problem was that the bottom of the function is never zero, so we figured that the domain was all numbers. But then we didn't know what to try in the function to get the range.

From the graph, the range is all numbers between 50 and 999 (approx.).

Description of Graph in Words

The graph starts off with a height of 50. It then begins to increase in a concave up fashion. Somewhere between 40 and 50 months, the graph stops being concave up and becomes concave down. It still keeps increasing, but now it begins to level off at a height of about 1000. The graph stays at the height of 1000 for the rest of the time.

Group Homework Problem : Section 1.4 #23 (Page 46)

Example   In month $t = 0$, a small group of rabbits escapes from a ship onto an island where there are no rabbits. A study determines that $p(t)$, the island rabbit population in month t, is given by :

$$p(t) = \frac{1000}{1 + 19(0.9)^t} \qquad , t \geq 0$$

(a)   Evaluate the expressions $p(0)$, $p(10)$, $p(50)$, and explain their meaning in the context of rabbits.

(b)   Sketch a graph of $p(t)$, $0 \leq t \leq 100$, and describe the graph in words. Does it suggest the kind of growth you would expect among rabbits on an island ?

(c)   Find $p^{-1}(50)$ and explain its meaning in terms of rabbits.

(d)   Estimate the range of $p(t)$. What does this tell you about the rabbit population ?

(e)   Explain how you would find the range of p algebraically.

Solution   (a)   The expressions that you have to evaluate are as follows :

$p(0) = 50$ This represents the number of rabbits that arrived on the island after having escaped from the ship.

$p(10) \approx 131$ This represents the number of rabbits inhabiting the island 10 months after the original 50 rabbits arrived on the island.

$p(50) \approx 911$ This represents the number of rabbits inhabiting the island 50 months after the original 50 rabbits arrived on the island.

(b)   The graph of the rabbit population of the island looks like :

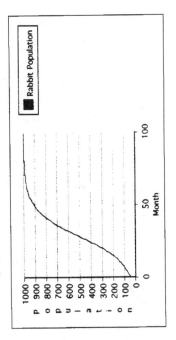

When the rabbits first escape onto the island, there are only 50 of them. I would expect that, at least initially, the rabbit population would grow quite slowly. Several factors contributing to this slow initial growth :

• There are simply not many rabbits to breed. The best scenario (at least as far as fast breeding is concerned) would be if all 50 rabbits that escaped were sexually mature, and had mainly females with just a few males. We are not told whether this is the case or not, but it is probably safe to say that the initial population was worse off than this.

• The rabbits would probably have spread out over the island. It seems reasonable to postulate that the frequency with which rabbits bred is roughly proportional to the frequency with which they met. So, initially, the rabbit population may have been quite spread out, not meeting each other very often, hence not breeding very often.

As the rabbit population grew, I would expect its rate of increase to accelerate, as is shown by the graph. This is because there would now be more rabbits per square mile on the island, and they would meet each other more frequently, and so breed more frequently. Also, since there are simply more rabbits having offspring, the rabbit population produces more offspring per unit time.

Finally, since the rabbits inhabit an island, which has only a certain amount of natural resources such as land area, food and water, only a certain number of rabbits can be sustained indefinitely. As the rabbit population grows, competition amongst rabbits for the natural resources they need to survive will become more and more intense. Eventually, not all the rabbits will not be able to get sufficient quantities of the natural resources that they need to breed, so the growth of the population will slow down, and settle (so the theory goes) at a population level (called the carrying capacity) where the birth rate balances the death rate, and there is no net population growth. This phase is represented by the graph flattening out.

(c) $p^{-1}(50) = 0$ This is the month in which the rabbit population of the island was 50 rabbits. This is because the population started at 50, and then rose thereafter.

(d)   Judging from the graph of the rabbit population, the range of $p(t)$ is :
$$50 \leq p(t) \leq 1000$$

This tells us that the maximum size that the rabbit population of the island will ever grow to is 1000 rabbits, whilst it will never drop below 50 rabbits. This means that the rabbit population will never die out.

(e)   Because the graph is always increasing for every x–value, the lowest possible y–value will be $p(0) = 50$, since $x = 0$ is the smallest possible x–value. The largest possible y–value will correspond to $x \rightarrow \infty$, which is 1000.

Example Problem: Students are expected to produce a longer, written report [2]

The problem described below is the first assignment for the first laboratory session in the Calculus 1 course (Math 31L) at Duke University. Students read a preamble describing a study of breast cancer among women. The study compared two groups: a case group (consisting of women who had been diagnosed with breast cancer), and a control group (consisting of women picked at random from the population). The preamble defined a quantity called the *raw relative risk* as the ratio of the percentage of the case group with a certain quality (for example, exercising three hours per week) to the percentage of the control group with the same quality.

Students were presented with a table of data, and a plot of this data (referred to as Figure 1). Students were asked to work through the questions listed below in groups of three or four, and then to write a report describing their findings.

Three specimens of students' work are included immediately after the questions that the students answered.

1. If 100 women in the control group exercise three hours a week, how many women in the case group exercise three hours per week?

2. In order to model the relationship between exercise and breast cancer we wish to find a line that is a *good fit* to the data. In your opinion, what characteristics of a line would make it a good fit?

3. Draw a straight line in Figure 1 that fits the data well (you may not use the statistical features of your calculator). Denote the relative risk by the variable R, and hours of exercise by the variable E. Determine the slope of the line you drew. Also determine the R intercept (where the line intersects the vertical axis). What is the equation of the line?

4. What does the R intercept represent in terms of exercise and the risk of breast cancer?

5. Is your R intercept the same as the data point that is on the R axis? Do you believe that the data point is where the "correct" model would actually have an intercept? Why or why not?

6. The slope of your line should be negative. Describe what this means in terms of exercise and the relative risk of breast cancer.

7. Use your linear model to estimate the relative risk of breast cancer for a woman who exercises 20 hours a week. Is this a reasonable estimate? We should, of course, qualify the sentence above by restricting to a certain set of values of E. Determine a set of values of E for which you believe your linear model will be valid, and complete the following sentence: "If a woman exercises between _____ and _____ hours per week, and if she increases her exercise by one hour per week, then her relative risk of breast cancer will be _____"

8. Estimate the relative risk of breast cancer if a woman exercises for 2.3 hours a week.

9. Estimate the relative risk of breast cancer if a woman exercises for five hours a week.

10. Find a value of E for which $R \approx 1.0$. Explain the significance of this value of E.

11. Actually, we do have another piece of data from this study: if a woman exercises more than five hours a week then (according to the study) her relative risk jumps to 1.2. Do you wish to refine the set of values for E for which the linear model appears valid? Suggest two possible reasons for why the data shows an increased relative risk of breast cancer when a woman exercises more than five hours per week.

12. In statistics, a common criterion for how well a line fits a set of data points is the sum of squares of the vertical distances from the line to the data points. Compute this sum for your line. Draw a different line that fits reasonably well and compute the sum for that line.

---

[2]From *Duke University Laboratory Calculus Manual, 1999–2000.*, [10].

Breast Cancer Lab

The goal of our lab was to find a linear line that best fit the four points that were given to us. On our first attempt we came up with an equation of (R= -.08E + .97). To get the slope of this equation we used the points (3.0, .73) and (4.0, .65). We used $(R_2-R_1)/(E_2-E_1)$ to find the slope of -.08. Then to find the R-intercept we plugged in the point (3.0, .73) into the equation R= -.08E + C. This yielded our R-intercept of .97. In reality this point on the graph would represent the relative risk of woman who do no exercise. However, this point is unrealistic because this point is not above the relative risk of the control group.

Another part of our lab asked us to estimate the risk for a woman who exercised 20 hours a week. Upon calculating this using our equation, we found that our equation yielded an unreasonable answer of –63% relative risk, which is obviously not possible in reality. Our group decided that our domain should be from 0 to 7 because any further and we would doubt the reliability of the graph. Later information told us that 0 to 5 should be the proper domain. We were also asked to estimate the relative risk of a woman who exercised 2.3 hours per week. Her risk was lowered by 21.4% compared to the control group. We calculated this answer using the equation given in the prior paragraph. Finally we were asked to compute the relative risk of a woman who exercised for 5 hours a week. We found that the risk was decreased by 43%, again compared to the control group.

In part L we were given a new piece of information that stated that if a woman exercised more than five hours a week her relative risk jumps to 1.2 or 20% higher than the control group. With this information we changed our domain to 0 to 5 as we stated earlier.

The final objective of part one of the lab was to test the accuracy of our equation. To do this we were given a formula of which the sum of the squares of the vertical distances from the line to the data points from which the equation is derived represents the accuracy of that line. We tested our line and found it to have an error of .0104. As a result of this we decided to see if we could find a line that would be more accurate. To do this we used the points (3.0, .73) and (0.0, 1.07). These points yielded a line equation of (R= -.113E + 1.07) and an error factor of .003965. Therefore the second line was a better match for the data points than our first line.

Risk Factors for Breast Cancer

Purpose: The purpose of this lab is to use information about risk factors for breast
              Cancer to mathematically develop best-fit lines to approximate the degree
              that different variables relate to the occurrence of breast cancer. These
              lines were then used to approximate values not given on original data
              table. Results were used for interpretation.

Procedure: See Duke University <u>Laboratory Calculus</u> Workbook, Lewis Blake and
              Michael Reed, 1998. Lab 1, page 7.

Results: a) A raw relative risk of .73 represents the ratio of women in the case

                 group to the women in the control group who exercise three hours per

                 week.

              b) A "good fit" line will come closest to all of the data points on the

                 graph.

              c) See attached.

              d) The R intercept represents the raw relative risk associated with zero

                 hours of exercise.

              e) Our R intercept was not the same as the data point of the R-axis. The

                 correct model would have an intercept at the data point R because the

                 "correct" model would include every data point. The inconsistency

                 resulted from the fact that our graph is a straight-line estimation.

              f) The negative slope means that the raw relative risk of breast cancer

                 decreases as hours of exercise per week increase.

g) The slope of our line is -.06. "If a woman increases her exercise by one hour per week, then her relative risk of breast cancer will be decreased by approximately six percent."

h) According to the equation of our linear model, a woman who exercises 20 hours per week would have a negative raw relative risk of contracting breast cancer. This is not a reasonable estimate. "If a woman exercises between zero and fifteen hours per week, and if she increases her exercise by one hour per week, then her relative risk of breast cancer will be decreased by approximately six percent with each hour."

i) If a woman 2.3 hours per week, her approximate relative risk would be .77.

j) If a woman exercises for five hours per week, her approximate relative risk would be .61.

k) R is approximately one at -1.5 hours of exercise per week. This value represents the point when there is an equal number of people in both the control and case groups.

l) Given the additional information, "if a woman exercises between zero and five hours per week, then her relative risk of breast cancer will be decreased by approximately six percent with each hour." One possible reasons that relative risk increases with more than five hours of exercise per week is that the women's bodies are more fatigued, and the immune system is not as strong.

Math Lab 1: Risk Factors for Breast Cancer

Utilizing the data provided by the study, we concluded that an increase in the amount of exercise per week decreases a woman's risk of breast cancer. For example, a woman who exercises 3 hours per week has a raw relative risk of 0.73. This means that a

HOURS OF EXERCISE VS. RELATIVE RISK

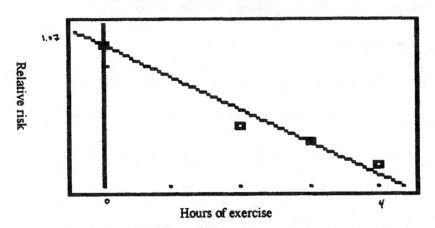

Hours of exercise

woman who exercises 3 hours per week is 27% less likely to develop breast cancer than the average woman is. We plotted the 4 data points, and then calculated the line of best fit. We estimated the line of best fit by drawing the line as close as possible to all points on the graph. We used (0, 1.07) and (3, 0.73) because our line crossed those points. Then, using the Difference Quotient, we calculated the slope of our line. We found the slope to be -0.113. It is a negative slope, which indicates that as the amount of exercise increases, the raw relative risk decreases. This relationship demonstrates an inversely proportional association between exercise and the risk of breast cancer. Using the equation for a line, y = mx +b, we were able to formulate the equation for the line to be y = -0.113x + 1.07. Once we established the equation of the line, we are able to extrapolate

points using this line. Thus, given the amount of time a woman spends exercising, we can determine how much her risk of getting cancer decreases.

Exploring the graph, we find that if a woman exercises 20 hours per week she would have a raw relative risk of −1.19, which is unrealistic. This shows the limitations of this model. The graph intercepts the R-axis at 1.07. This represents the relative risk of developing breast cancer if you do not exercise. A woman who never exercises is 7 % more likely to develop breast cancer. We understand, however, that the "correct" line of best fit would not actually hit the Y-axis at exactly 1.07, because the actual line of best fit will not hit every point. The correct model will come as close as possible to as many points as possible on the graph.

Using this model we can predict the raw relative risk using the number of hours of exercise. If a woman exercises 2.3 hours, we can determine her relative risk of developing breast cancer. By substituting 2.3 for x into the equation $y = -0.113x + 1.07$ we can conclude that her relative risk becomes 0.81. We can also go the other way: if the relative risk is known, we can find the number of hours of exercise. If we want to find out how much the average women in the control group exercises, we can substitute 1.0 in for y into the equation $y = -0.113x + 1.07$. We concluded that the average woman in the control group exercises 0.62 hours per week.

This model does have some limitations. We noticed the limitations when we found from new data that the relative risk for a woman who exercises more then 5 hours increases to 1.2. The reason for this is unclear, but it may represent that whatever benefit gained from exercise as a means of preventing or decreasing the incidence of cancer has been surpassed. For example, if the mechanism by which exercise protects a woman from breast cancer is by decreasing body fat content, then there will come a point at which further exercise will no longer decrease fat content. It may even become harmful to the women and actually be a factor in increasing her risk of breast cancer. We concluded that our model is only valid for women who exercise between 0-4 hours per week.

A Few Suggestions for Student Feedback

Adapted from suggestions on grading by Bob Megginson and from suggestions on grading writing developed in cooperation with William Hoese.

Although "feedback to students" is often equated with written comments on students' work, you should bear in mind that the numbers you assign to students' work are also a form of feedback. The suggestions offered here address both written comments and numerical scores.

1. Students' complaints about "nit-picking" usually are caused by the instructor's littering the problem with - 1/2's and -1's for individual errors. It is usually far better to read the entire exercise (or part of the exercise, if it is divided into parts (a), (b), and so forth), correcting errors as you go, and then assign one number as the grade for the entire unit consisting of the exercise or part. In short, you should *correct* all errors, no matter how small, but the smallest unit to which a *number*, either positive or negative, should be attached is an exercise or a part of an exercise. (Now see the next suggestion about negative numbers on grading.)

2. Students often seem to react better when you give them points rather than take them away. If an exercise is worth 8 points and the student misses two of those points, it seems to be better to write 6 beside the exercise rather than -2. Also, using a "give" rather than a "take away" system can lead to better grading, since the instructor is more likely to think of what the effort is worth overall rather than which individual errors require deductions.

3. Just as you would not shout disapproval at students in class, you should not shout disapproval on paper. In comments that could be construed to be negative, you should avoid such emphatic devices as underlining, capitalizing and exclamation marks, since these are just written forms of shouting that can hurt students' feelings. Put yourself in the students' shoes when reading the following two remarks:

   - This problem cannot be done this way since the function is not linear.

   - This problem **CANNOT** be done this way since the function is **NOT LINEAR!!!**

   The first is a statement of fact, and the student should not take this amiss. The second says, essentially, that the instructor considers the approach (and the student taking the approach) to be particularly misguided. It may be true that the instructor does consider the approach to be so, but nothing is to be gained by saying that. Pointing out the error is sufficient. So what about the following compromise?

   - This problem cannot be done this way since the function is obviously not linear.

   This is perhaps not quite so bad, but what is really added by the word "obviously?" The instructor is effectively saying that the student is somehow deficient for not noticing something that should be "obvious." *In general, avoid the use of emphatic devices such as underlining, all capitals, exclamation points, and emphatic adjectives when correcting errors.* Of course, it is perfectly fine to use those devices in *positive* comments!

4. The preceding item leads to the question of what to do when the student really is bluffing and what is written is nonsense. While the temptation is strong to retaliate, the point is adequately conveyed by a zero grade and a neutral comment such as "This is incorrect—see the examples given in class."

5. Some students do not realize that they need to process comments and utilize suggestions in order to improve their writing when they compose new reports. One possible solution is to make some of the assignments "double submission" assignments where the first draft of the paper is commented by the instructor, and the second draft is graded. Along with the second draft, the students are required to submit a description of how they addressed each of the issues that the instructor commented on. (This is much like the process of submitting and revising a paper for submission to a professional journal.)

6. Students often read comments through their paper just like a list (instead of reading their text that the comment addresses). It is almost as if students think the comments themselves are linked together and comprise a "mini-paper," rather than closely examining each of the comments (and the relevant section of their text) individually. Although a "double submission" policy can make more work for the instructor, it can encourage students to work on each of your comments individually. A compromise is to make only the first one or two assignments double submission.

7. Suggestions and comments are often more effective when they ask a question and are written as complete sentences. Single word comments are often misunderstood or not viewed as serious by students.

8. Although you may grade on a numerical scale, don't feel as though you *have* to use every number on that scale. For example, one collection of "categories" of papers and their corresponding numerical scores is given below.

   - Excellent paper – 20/20.
   - "Good" paper – 17/20.
   - "Needs Improvement" – 14/20.
   - "Poor" – $\leq 12/20$.

9. *To the extent to which this is possible*, evaluate students' success at separate goals separately. For example, it may help to assess the "writing" aspects and "content" aspects of the paper separately. Some points that could be assessed in each of these two areas are listed below.

   - Presentation, communication skills, language skills.
     - Relevant issues are addressed in a direct fashion.
     - Central ideas are communicated effectively.
     - Arguments and discussion are presented in a logical order.
     - Style and grammar are acceptable.
   - Evidence of conceptual understanding.
     - Demonstrates a clear understanding of important issues.
     - Demonstrates an awareness and good understanding of the implications of data and information presented.
     - Supports conclusions and arguments with relevant references.
     - Makes connections between data, references and other relevant information.
     - Draws conclusions that follow from the information and arguments presented.

10. A summary of good points and points of improvement at the end of the paper can be very helpful for students, especially if there are systemic problems with the paper, or with the student's style of writing. These summaries can be difficult and time-consuming to write, however.

11. Try to remember that students need both positive feedback and constructive criticism for their papers. They need to know what they've done right as well as where they can improve.

Grading Rubrics

Grading rubrics (or grading schemes) are an attempt by an instructor to create a system for assigning points to students' work. Grading rubrics are popular because they help to ensure consistency in what the instructor awards points for, how many points are awarded, and because they allow instructors to do a lot of their serious thinking in a batch when they are still "fresh" at the beginning of the grading session. Sometimes, if there is a lot of student work to be graded, instructors can become mentally fatigued, which may affect their thinking.

In this handout, we will describe two different kinds of grading rubrics: generic rubrics and more specific rubrics.

1. **Generic Rubrics**

   General rubrics are very generic grading schemes that apply to grading almost any piece of student work. The consistent and fair use of a generic rubric depends very heavily on the individual instructor's ability to gain a general sense of the "quality" of each piece of student work.

   A generic rubric can be a way to grade very quickly, because the instructor is primarily concerned with establishing a general "sense" of each student's work without becoming obsessively concerned with awarding or removing points based on every single detail of the student's work.

   The careless use of a generic rubric can provide students with very little useful feedback on their work. For example, if the instructor meticulously assigns points for every detail of the students' work, then students can find out what they did wrong by looking for places where they did not get any credit. If all that students have to go on is "3/5," then it can be difficult for them to see how to improve, and most importantly, what mathematical concepts to study harder. Using a generic rubric demands that the instructor carefully comment on students' work in addition to deciding on a numerical score.

   Generic rubrics are commonly described by a $0 - 5$ point scale, with the somewhat obvious advice that "good work gets a 5, reasonable work gets a 3 and unacceptable work gets a 1."

   An example of a generic rubric is given below[3].

   5: Not necessarily "perfect." Some write-up problems are okay. A couple of mathematical equivalents of "typos" are okay. An intelligent, thoughtful, but incorrect approach may get this score, too. In general, a (possibly flawed) work of art.

   3: Serious write-up problems (although some effort shown) with good mathematics. Or a seriously flawed approach written up well. Or a combination thereof. A "3" problem has the feeling of, "Look, let's just write up what we have so far and move on."

   1: This means some minimal effort was made, but the student does not have the concept. If someone just writes down the right number, but shows minimal work, then this may be the appropriate grade. Likewise, if a student wrote gibberish, but puts effort into the write-up, then this may be the appropriate grade.

   0: This grade is reserved for situations where the students show absolutely **no** work whatsoever, or do not turn anything in.

   A score of 4 is assigned when it is very difficult to decide between a 3 and a 5, and the score of 2 used similarly.[4]

   - **Advantages of Generic Rubrics**

     - There is usually little work that needs to be done to create the rubric.
     - Applies to many different kinds of assignments.
     - Numerical scores can be assigned very quickly.

---

[3]Adapted from "How Doug Marks Papers" by Douglas Shaw

[4]In the original version of this generic rubric, Doug Shaw described his thought process as he assigned scores to students' work: "I ask myself, 'Five, three or one?' and I have two "in–between" grades if I can't quickly decide."

  – The instructor can first correct and comment on all of the students' work, and then go through
    and assign numerical scores to all of the students' work. Doing the same thing to all of the
    work at once can help to improve consistency.

- **Disadvantages of Generic Rubrics**

  – Relies very heavily on instructor's ability to intuitively develop a sense of the "quality" of
    students' work.
  – The points assigned are not a definite guide for students on where they need to improve.
  – If the instructor does not write comments on the paper, then students do not get any specific
    feedback on their work.
  – The instructor's judgment may change over time (e.g., due to fatigue during long sessions)
    which may affect the number of points the instructor awards to student work.
  – The use of a generic rubric by a grader who is not conscientious may make it possible for the
    grader to assess student work in a superficial way.

2. **More Specific Rubrics**

Grading rubrics do not need to apply to a wide range of assignments or problems. Rubrics can be
developed that apply to a particular style of assignment (or style of assignment, such as an expository
writing assignment). Rubrics can be even more specialized, applying to only one assignment, or even
one problem on one assignment. This is the kind of rubric that we describe here.

Specific rubrics are an excellent place for new instructors to begin, because they are usually very
concrete and explicit on what points are awarded or deducted for. A process for creating one of these
more specific rubrics is illustrated in Figure 1.

Some instructors like to begin by reviewing some examples of students' work, so that they can see how
students have interpreted the question, and how they are attempting to respond to it. This preliminary
step can save a lot of headaches if many students have interpreted the problem in quite a different (but
valid) way from how you intended the students to interpret the problem. Being aware of this from the
outset can help you to design a grading rubric that takes these alternative interpretations into account.

The "looking back" step depicted in Figure 1 involves taking a moment or two to review the rubric
that you have created, and to make sure that your sense of what is important in the problem has not
been "lost in the details." Another important point to consider at this point is how "just" the rubric
is. For example, if a student makes a small error at one point, but otherwise does an excellent job, will
the one small error lower the score a lot?

An example of a specific grading rubric is given in Figure 2.

- **Advantages of Specific Rubrics**

  – Usually very concrete and straight forward to use.
  – Can help to preserve high standards for you to hold students to.
  – You do most of your thinking about what you will give points for while you are still fresh at
    the beginning of the grading session.
  – Helps to ensure consistency.
  – Can help grading to go quickly.

- **Disadvantages of Specific Rubrics**

  – You still have to figure out what you will do about unusual student answers so that you treat
    them fairly.
  – If students interpret the problem very differently from what you intend, then you may have
    to make more than one grading rubric.
  – These kinds of rubrics can be unjust if applied too rigidly.
  – There can be quite a lot of work involved in figuring out and writing the rubric in the first
    place.

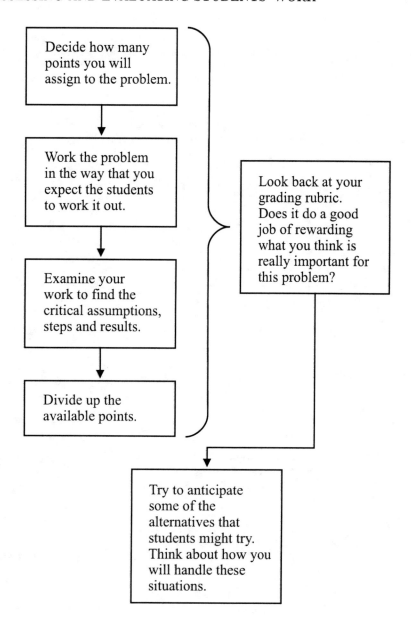

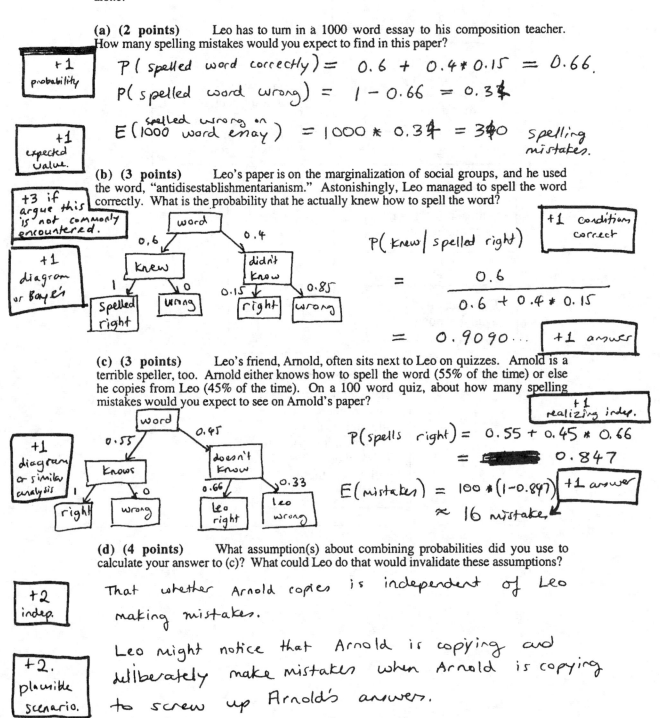

1. Leo is not a very good speller - in fact, his composition teacher estimates that Leo only knows how to spell about 60% of commonly encountered English words. When Leo was young, he took the "Hooked on Phonics" course. If he encounters a word that he does not know how to spell, he has a 15% chance of getting it right, based on phonics alone.

(a) (2 points) Leo has to turn in a 1000 word essay to his composition teacher. How many spelling mistakes would you expect to find in this paper?

+1 probability

$$P(\text{spelled word correctly}) = 0.6 + 0.4 * 0.15 = 0.66$$
$$P(\text{spelled word wrong}) = 1 - 0.66 = 0.34$$

+1 expected value.

$$E\left(\begin{smallmatrix}\text{spelled wrong on}\\ 1000 \text{ word essay}\end{smallmatrix}\right) = 1000 * 0.34 = 340 \text{ spelling mistakes.}$$

(b) (3 points) Leo's paper is on the marginalization of social groups, and he used the word, "antidisestablishmentarianism." Astonishingly, Leo managed to spell the word correctly. What is the probability that he actually knew how to spell the word?

+3 if argue this is not commonly encountered.

+1 diagram or Baye's

+1 condition correct

$$P(\text{knew} \mid \text{spelled right})$$
$$= \frac{0.6}{0.6 + 0.4 * 0.15}$$
$$= 0.9090...$$ +1 answer

(c) (3 points) Leo's friend, Arnold, often sits next to Leo on quizzes. Arnold is a terrible speller, too. Arnold either knows how to spell the word (55% of the time) or else he copies from Leo (45% of the time). On a 100 word quiz, about how many spelling mistakes would you expect to see on Arnold's paper?

+1 diagram or similar analysis

+1 realizing indep.

$$P(\text{spells right}) = 0.55 + 0.45 * 0.66 = 0.847$$
$$E(\text{mistakes}) = 100 * (1 - 0.847) \approx 16 \text{ mistakes}$$ +1 answer

(d) (4 points) What assumption(s) about combining probabilities did you use to calculate your answer to (c)? What could Leo do that would invalidate these assumptions?

+2 indep.

That whether Arnold copies is independent of Leo making mistakes.

+2. plausible scenario.

Leo might notice that Arnold is copying and deliberately make mistakes when Arnold is copying to screw up Arnold's answers.

Figure 2. A specific grading rubric.

Assessment Issues for Student Writing

1. Reading, commenting and evaluating writing can be a very time-consuming process, especially when you are new at it.

2. How much do you allow yourself to "read the students' minds" when you read their writing? For example, have students failed to mention their assumptions because they consider these to be obvious, or do they not understand the problem well enough to have realized what assumptions they have made? Are comments by students inarticulate or incomplete because students understand the mathematics, but are not accustomed to writing, or do students simply not understand the mathematics clearly enough to describe the math coherently? It should be noted that this same concern should also apply to more traditional, symbol-manipulation type assignments.

3. What do you assess and grade? Many mathematicians feel that while they are very well qualified to evaluate the correctness and appropriateness of students' mathematical manipulations, they feel much less qualified to evaluate students' ability to:

   - spell words correctly,
   - use grammar and punctuation correctly,
   - construct paragraphs that support the students' discussion,
   - write using an appropriate register and sense of style,
   - organize their work on the page to help readers easily follow the discussion, and so on.

4. Plagiarism. You may need to consider the possibility of plagiarism in several different ways.

   - **Failure to properly reference quoted material.** Some instructors ask their students to complete assignments where the students have to look up and learn about theorems or concepts. Students are typically asked to write some kind of report describing what they have learned. The problem of plagiarism is clear here: some students may quote material from their sources without properly citing or referencing their source.

   - **Extent of collaboration on team homework and projects.** Students working in teams are naturally encouraged (sometimes required) to seek out help from other members of their team. This spirit of collaboration can sometimes extend a little further than desired. For example, students may assume that they may seek help from people who are in other teams. While this may not be an issue, it is important for instructors to be very clear on whether this is appropriate or not.

   - **Word processors.** Many instructors insist that students write their reports using word processors. Because it is very easy for students to share their files, it becomes very difficult for the instructor to decide exactly who has done the write-up.

   We note that instances of blatant, large-scale plagiarism are (in our experience) rare.

5. Consistency (both from student to student on a given assignment, but also from assignment to assignment).

6.7   Suggested Reading

- Annalisa Crannell. "How To Grade 300 Mathematical Essays and Live to Tell the Tale." [34]

  This article describes a system for grading writing assignments according to an eleven point checklist. This system is also mentioned in the book by Meier and Rishel, although they suggest alternative approaches as well. According to Crannell, an advantage of this system is that it is "time-efficient." Another advantage, especially for beginning instructors with little experience with assessing and grading writing, is that the checklist is fairly concrete.

- George Gopen and David Smith. "What's an Assignment Like You Doing in a Course Like This?: Writing to Learn Mathematics." [56]

  This article encompasses a wide range of issues in the use of writing assignments in mathematics classes. The article begins by describing a calculus course taught by David Smith involving a computer laboratory component and weekly writing assignments. The article describes the dissociation of numbers and words, and lists some common problems that students have with writing about mathematics. The main problems listed are: (1) difficulty finding conceptual rather than factual content, (2) failure to connect narrative with data, (3) failure to make connection explicit, and (4) denial or suppression of mistakes. As responses to these problems, the authors describe "Reader Expectation Theory," a methodology for analyzing and teaching writing. The article includes numerous illustrations of this methodology at work. The authors also advocate a "double submission" system, where the instructor comments on a first draft of the assignment, the students revise their work, and the instructor grades the second version of the assignment. Two "before and after" samples are given to illustrate how much students can improve both their writing and their grasp of mathematical ideas using the double submission system. Finally, several procedures are listed that can encourage students to improve both their expression and understanding of mathematics.

- Wilbert McKeachie. "*Teaching Tips.*" [81]

  Chapter 6 (Tests and Examinations) lists seven procedures for grading essay questions. These procedures are intended to improve the reliability or consistency with which instructors grade essay questions. As McKeachie points out, all entail work, and some may be prohibitively labor-intensive. One point that may be helpful for beginners is the suggestion to choose examples of "excellent," "fair," and "poor" papers, and keep these aside as an aide to memory.

- John Meier and Thomas Rishel. "*Writing in the Teaching and Learning of Mathematics.* MAA Notes #48." [82]

  Chapter 6 deals with grading essays. The authors suggest developing a "checklist" for use when grading writing assignments, even if this is an internal, implicit checklist rather than an explicit written document. Meier and Rishel note that they often check to see if

  - the student has included a clear introduction,
  - the student has included a clear conclusion,
  - the essay is well organized and clear to follow,
  - the reader has to stop and say, "Wait! What's going on here?"
  - there is a clear indication as to why the student is performing his or her calculations,
  - the mathematical arguments are clear and well stated,
  - the paragraph structure supports the general discussion, and,
  - spelling, grammar and punctuation are correct.

- Andrew Sterrett, ed. "*Using Writing to Teach Mathematics.* MAA Notes #16." [109]

  This is a collection of articles describing individual teachers' experience of using writing assignments in mathematics classes. A very wide variety of programs are described in various levels of detail. Some of the accounts describe criteria that instructors used to assess and grade writing-intensive assignments.

Chapter 7

Managing Homework Teams

As we mentioned in Chapter 4, cooperative learning is gaining popularity as a pedagogical device in introductory collegiate mathematics courses. Instructors are encouraging students to learn in teams in or out of the classroom, and many instructors, ourselves included, have their students work cooperatively in *and* out of the classroom. Chapter 4 was primarily concerned with issues of the implementation of cooperative learning in the classroom, whereas in this chapter we explore issues of implementation for cooperative learning outside of the classroom.

In this chapter we discuss the management of "homework teams," because we require our students to meet regularly outside of class to work on difficult, non-routine homework problems. Other instructors require their students to meet in teams outside of class to work on long-term projects, prepare presentations, do laboratory activities, etc. The issues discussed in this chapter arise in any setting where team efforts are required, especially outside of the watchful eye of the instructor. This meeting is easily adapted to address the kinds of activities mentioned above, so as you read consider the phrase "homework teams" to be more broadly defined.

In the introduction to Chapter 4, we pointed out that cooperative learning can have positive or negative effects, depending on how it is implemented. Much of the literature on the implementation of cooperative learning in a college mathematics class is connected with the implementation of cooperative learning in a classroom environment where the instructor is physically present. In this case the instructor can take some kind of executive role in helping students to direct their energies productively, and to foster the kinds of interactions that may lead to cognitive growth and meaningful understanding of the concepts of mathematics.

"Homework teams" present a different kind of challenge. While the instructor can model helpful cooperative behavior and reward students who exhibit this behavior during class, the instructor is simply not present to execute this function during the homework teams' meetings. Nevertheless, many students still expect the instructor to be able to remedy problem situations that arise with "homework teams."

In this meeting we first address the issue of forming homework teams. Some problem issues can be circumvented by a thoughtful approach to forming the teams. An approach to the initial formation of homework teams was given in Block B of the pre-semester orientation found in Chapter 3. Some practitioners delay the formation of teams until somewhat later in the semester, at which time more information is available for use when formulating the teams. Other practitioners regroup their teams periodically throughout a semester. For these reasons, in this chapter we explore in more depth some options for grouping teams.

We then endeavor to describe some of the more common kinds of problems that teams seem to experience. We try to trace these problems to possible root causes. Finally, we suggest ways in which instructors can try to address the difficulties that many homework teams experience.

7.1 Description and Purpose of the Meeting

This meeting is intended to be a fifty minute session. This meeting can be used as a venue for distributing written information to participants, although this can also be done in advance of the meeting. The main purpose of this meeting is to bring less experienced instructors into contact with instructors who are experienced users of homework teams. For this reason, a substantial proportion of the meeting is devoted to somewhat free-form discussions. During these sessions, the role of the meeting leader is not so much to provide directive leadership, but to facilitate the discussion, to prevent individuals from dominating, and to make sure that the discussion does not get diverted to other subjects.

7.2 Goals for the Meeting

1. To provide instructors with guidelines for implementing various strategies for selecting members for homework teams.

2. To provide experienced instructors with a forum to share the important lessons that they feel they have learned in managing homework teams.

3. To alert instructors to some of the potential difficulties with homework teams, and to initiate a discussion with experienced instructors on ways to manage these difficulties.

4. To provide instructors with guidelines for identifying, confronting, and mediating conflicts between homework team members.

5. To provide instructors with a forum for hearing others' thoughts on the issues outlined above.

7.3 Preparation for the Meeting

1. If time permits, and if people in the department are cooperative, take some time to speak with experienced instructors concerning their experiences with managing homework teams. Questions that could be used to focus the discussion include the following.

   - What do you feel was the single biggest challenge that you faced when using homework teams?
   - In your opinion, what is the greatest hurdle that students face when trying to work together on a team?
   - Speaking as the person who administers and grades the homework, how is team homework different from individual homework?
   - If you could change one thing about the way that team homework is handled in our department, what would that be?
   - What advantages can team homework present over individual homework?
   - What is your sense of how your students have felt about doing team homework? (If the answer is negative, you could follow up with a question like, "What ideas do you have for addressing the students' concerns?")

2. Review the list of strategies for forming homework teams. Add new strategies that you have used, or that experienced instructors have told you about, to the list. Add problems that are specifically concerned with the actual *formation* of homework teams. Problems that occur when teams are formed and have been working together for some time should be reserved for the "Diagnosis and Treatment of Sick Groups" handout.

3. Review the list of potential problems with homework teams on the handout, "Diagnosis and Treatment of Sick Groups." Create new lists—perhaps using the same format of problem, cause, and suggestions for treatment—including problems that are endemic to your department, or which are important but not included on the existing list.

4. Prepare some kind of ice-breaker or focusing activity for each discussion that you plan to initiate. Ice-breakers that have worked include "Tricky Situation" sheets, that describe incidents touching on issues relevant to the topic of the discussion. For examples, see the "Meeting Materials" section of this chapter.

5. Circulate the agenda for the meeting to all participants well in advance.

6. Make copies of meeting materials, and any questionnaires that you plan to distribute, and confirm attendance with participants.

7.4 Agenda for the Meeting (to be distributed to participants)

1. Review and description of methods for forming homework teams. There will be an opportunity to discuss the merits of different strategies with experienced instructors.

2. Discussion of some of the problems that homework teams experience. There will be an opportunity to discuss potential problems, and potential solutions, with experienced instructors.

3. Outline of your responsibilities to homework teams. In particular, an indication of what kind of problems you may or may not want to involve yourself in. We will also talk about some strategies for resolving conflicts in teams. There will be an opportunity for experienced instructors to relate the experiences that they have had with teams.

4. A discussion of some of the ways of measuring team performance.

7.5 Outline for Running the Meeting

:01–:05 Introduce the meeting, and briefly review the agenda.

:05–:15 Distribute copies of the handout entitled "Forming Homework Teams." If it is standard practice in your department to change homework teams periodically throughout a semester, then point out to the instructors that there are methods that they can use to do this. If it is standard practice in your department to use semester-long homework teams, and if these teams have already been formed, either skip this section of the meeting, or briefly use it to give the instructors ideas for forming homework teams in future semesters.

Give participants a chance to look over the different strategies. There are several ways that experienced instructors could become involved to lead a discussion on their experiences using the different strategies. One possibility is to have the experienced instructors form a panel at the front of the meeting room. One after another, the experienced instructors could relate the particulars of their implementations of the various methods. The meeting leader could act as a moderator in this situation, and try to get the experienced instructors to mention which of the strategies they have enjoyed particular success with, and what they feel contributed to that success. The meeting leader could conclude this portion of the meeting by inviting participants to ask questions of the panelists. Some particular points that may come out of this discussion are

- Should teams be heterogeneous or homogeneous? Should they be heterogeneous/homogeneous according to
  - test scores?
  - race?
  - gender?
  - mathematical experience or background?
  - major or intended career path?
- Should students have some say in how their groups are formed? If so,
  - should students decide how big teams should be?
  - should students have complete authority to pick their team mates?
  - what are you going to do about students who have no preferences?

Another possibility for running this part of the meeting is to break the group of participants up into small groups of three to six people per group. Each small group could be deliberately seeded with one or two experienced instructors who will relate their experiences to the other participants. To help beginning instructors hear many viewpoints, this part of the meeting could be ended by calling for the groups' attention to be directed to the front of the room, and having the experienced instructors briefly report the strategy that they have enjoyed the most success with, and what they feel has contributed to that success, to all participants at the meeting.

:15–:25 Do an ice-breaker activity for a discussion of problems with forming homework teams. One option here is to make and distribute copies of the handout "Tricky Situations with Forming Teams" given in the "Meeting Materials" section of this chapter. Give the participants a minute or two to look the sheet over. Ask the participants if there is anything that strikes a particular chord with them, or anything that they see as being a significant and potential problem. As people begin to speak up, ask for solutions from the other meeting participants, especially the experienced instructors. When wrapping up this segment of the meeting, point out that dealing with these kinds of issues represents an ongoing problem, and identify resource people in the department that beginning instructors can contact if they find themselves having to deal with a problem.

:25–:35 Do a similar ice-breaker activity for a discussion of problems with managing homework teams. One option here is to make and distribute copies of the handout "Tricky Situations with Managing Teams." Again follow a time for reading and reflection with some group discussion of the problems and possible solutions.

Distribute the handout on the "Diagnosis and Treatment of Sick Groups," and give the participants a few minutes to look it over. Here, you could use a panel again, or you could break into groups. Alternatively, the meeting leader could work through the handout. This handout was developed in response to reported problems with homework teams in a calculus course. The specifics of this course that are relevant to these handouts are listed below.

- Students were assigned to teams with four members. Each member was required to assume one of four roles. The *manager* was in charge of ensuring that the homework was completed to a satisfactory standard and handed in on time. The *scribe* was in charge of writing up the final version of the homework assignment. The *reporter* wrote an account of the team's meeting, with the intention of describing any problems that the group encountered to the instructor. This report was stapled to the front of the homework before it was handed in. Finally, the *clarifier* was in charge of making sure that all team members understood the solutions to the homework.

- Students were required to buy a course pack that described these roles, and also gave some suggestions on making homework meetings more productive. For example, students were encouraged to at least read the problems before showing up to the meeting.

- The official course policy was that the roles that the students played should rotate after every assignment. The pragmatic reason for this was that the scribe's job usually involved a lot more work than the others, and the work should be shared among all team members. Some groups were reluctant to do this, often because they did not trust a particular group member (or members) to do a good job of writing up the assignment. Since students usually wrote their solution by hand, it was relatively easy for instructors to detect that the same person was acting as the scribe each time.

- Homework quizzes were a measure adopted by some instructors when they felt that not all of the team members were making an effort to understand the solutions to the team homework. These were in-class, timed quizzes that consisted of one of the problems from the team homework. Generally, students were informed beforehand of the quiz format, and in some cases the instructor had graded and returned the assignment that the quiz was based on before the students took the quiz.

Involve the experiences of other instructors as much as time permits. It is important to stress how the problem was detected, and what steps were taken to try to sort things out. During this segment of the meeting, the meeting leader could try to indicate—or even explicitly state, if there is a policy—what level of involvement classroom instructors are expected to maintain in helping homework teams to solve their non-mathematical problems.

:35–:45 Pass out the handout entitled "Team Evaluation Form." Give the participants a chance to read the items that are included on the evaluation form. Ask participants for ideas on how they could use the information from this evaluation form, and when in the semester such forms could be usefully distributed. Ask the participants if there is any information that they think important that is not included on the evaluation form. If any participants identify a valuable point, ask the participant to elaborate on why an instructor may be interested in that information, and how the instructor would make use of the information.

End the meeting on a positive note by passing out copies of the handout "Hints on How to Make Groups Work" provided in the "Meeting Materials" section of this chapter. Allow a brief time of discussion, where the participants can give any other observations on how to help teams run effectively.

:45–:50 Wrap up the meeting with a word of thanks, and (as appropriate) time for participants' questions, a summary, suggestions for additional readings, or a questionnaire.

7.6 Meeting Materials.

1. Copies of the handout entitled "Forming Homework Groups."

2. Copies of the handout entitled "Tricky Situations with Forming Teams."

3. Copies of the handout entitled "Tricky Situations with Managing Teams."

4. Copies of the handout entitled "Diagnosis and Treatment of Sick Groups."

5. Copies of the handout entitled "Hints on How to Make Groups Work."

6. Copies of the handout entitled "Team Evaluation Form."

Forming Homework Groups
Four Approaches

Adapted from suggestions kindly provided by
Robert Megginson, Marius Irgens and Doug McCulloch.

Students' Choice

In this scheme, every student submits a list of persons with whom the student would like to work, and a list of those persons with whom the student would **NOT** like to work. When doing this, it is important to stress that students may not get their first choices, due to the combinatorial difficulties of matching students with different opinions of each other.

Advantages

- The students like this method.

- The resulting groups tend to work well.

Disadvantages

- If used to make "new" teams, this method often results in the reconstruction of the very first homework groups.

- Some people may feel very bad if they are rejected by people with whom they would really like to work.

- Some people who do not get their first choice may feel that this is because you have something against them personally.

Implementation

- Have students hand you a note card with

    - their name (e.g., circled so that it stands out),
    - A list of n names of people with whom they would like to work,
    - A list of m names of people with whom they would not like not to work.

- Use these cards to organize people into groups as best you can. It is very unlikely that things will work out perfectly. In addition, some people will forget to submit cards, further complicating your considerations.

Mixtures of Ability

In this scheme, the instructor organizes the homework teams on the basis of the students' scores on uniform exams. When implementing such a scheme, it is important to adjust groups to compensate for obvious problems involving gender, race, or living area.

Advantages

- The groups tend to be uniform in their overall ability to do the exercises.

Disadvantages

- The strongest students do not get to work with students as strong as they are.

- The groups may mix people with very different levels of commitment to the course.

- Very strong students may dominate group meetings.

- Students who are able, but have not performed well due to a lack of personal commitment to the course, may try to leach off others who are committed to doing well in the course.

Implementation

- For the sake of argument, suppose that there are 32 people in the class, to be organized into eight groups (a to h) with four members each.

- List the students in order of exam scores (1 represents the lowest score, 32 the highest score).

- Form the groups as follows.

a.	1, 9, 24, 32
b.	2, 10, 23, 31
c.	3, 11, 22, 30
d.	4, 12, 21, 29
e.	5, 13, 20, 28
f.	6, 14, 19, 27
g.	7, 15, 18, 26
h.	8, 16, 17, 25

Homogeneity of Interest

In this scheme, the instructor assigns students to homework teams by considering common interests, such as majors. For example, the instructor may try to put engineering majors together, kinesiology majors together, and so forth.

Advantages

- The commonality of shared interests may give the people in the groups more of a base to communicate. Hopefully this would lead to better group work.

Disadvantages

- The commonality of interests other than mathematics may cause group meetings to lose focus as the group members spend more and more time discussing other issues.

Implementation

- Use information from student data sheets or mathematical autobiographies to arrange students in homework groups.

Living/Working Areas

In this scheme, the instructor assigns students to homework teams by asking students to indicate in which parts of campus or town they would like have their homework team based.

Advantages

- It is usually easier for students to get together when they live or work in the same area.

- If students have to travel to a group meeting, and they have another group member living nearby, they can travel to the meeting in pairs.

Disadvantages

- If this was the method used to initially form homework teams, then it will probably not be useful for making "new" teams.

- It is possible that geographically assigned groups will be unsatisfactory from another point of view, because of problems involving ability, commitment to the course, gender, etc.

Implementation

- Use information from student data sheets to group together those who live close to one another. Another possibility is to hand out a map of the campus area and have students indicate where on the map they would like to have their homework team based.

Commitment to the Course/Commitment to Group Learning

In this scheme, the instructor assigns students to homework teams by attempting to determine students' attitudes towards and commitment to both the course and the process of working in teams.

Advantages

- This method attempts to minimize mismatches of student expectations.

- Hard working students will likely not be matched with "lazy" students, thereby reducing the risk that a student's grade will be unfavorably affected by the irresponsibility of another student.

Advantages

- It can be difficult to accurately determine students' attitudes and commitments.

- Attitudes and commitments can influence group functioning, but group functioning can also influence attitudes and commitments. Therefore, students may change once they are put into their groups.

Implementation

- Gather information from a team evaluation form, your personal observations and conversations, and questions on the student data sheet such as "How do you feel about the prospect of working in groups?" Match students homogeneously according to their level of commitment to the course, their work ethic, and their ability to function well in, and contribute to, their teams.

Tricky Situations with Forming Teams

1. On the first day of class, you announce that you will be forming students into teams, and that they will have to meet to do their homework assignments outside of class. At the end of the hour, Jeremy approaches you. He tells you that he is a commuter, and he works full time. He asks if he can do the homework individually, because it would be very hard for him to meet with other students.

2. You collect data from students, and arrange them into teams. One day after announcing the teams, you receive an E-mail message from Jamie saying that her team cannot find a time when they can all meet.

3. After announcing a team homework policy in class, two students approach you. They are both trying to get into business school, and say they need a high grade in calculus to achieve this. They both express concern that their grades may be hurt if their team members are not that good, and ask if they can do the homework by themselves.

4. Your university has two campuses, which are a thirty minute bus ride apart. Most of the people in the class live near campus 'A,' but six live near campus 'B.' Your course director has announced a policy that teams should have a maximum of four members.

Tricky Situations with Managing Teams

1. Brett reports to you that Mary has not been attending team meetings. Meanwhile, Mary tells you that the rest of the group has purposefully been excluding her when planning meetings. You arrange to talk to the entire team. Brett insists that he has called Mary repeatedly. Mary insists that she has never received a call concerning when or where the meeting times are. The other two team members are reluctant to say anything, but they seem to be siding with Brett.

2. You have instructed the students to take turns writing up the homework. Steve's group always turns in a typed homework sheet, and they report that they are faithfully rotating who types up the work. Steve does exceptionally well on his tests, as does the group on their homework. The other group members are much weaker on tests. You begin to suspect that Steve is typing up all of the homework, and using the fact that it is not handwritten to make it difficult to detect that he is the one always doing it.

3. Carrie comes to your office to discuss the group work. Carrie is a very bright mathematics student, and she feels that meeting in the groups is a waste of her time, because she invariably spends the entire meeting explaining the material to the other students. She feels that she can learn the material well by herself, and that her time would be more efficiently spent without the group meetings.

4. Meredith calls you on the phone at your office. She is concerned that Tony does not have the same commitment to their team as the other members. She says that Tony does go to the team meetings, but he never does anything to participate. Meredith quotes Tony as saying, "Calculus is easy—I had it in high school."

5. At the drop deadline, Ross and Matthew, who are from the same group of four, drop the course. No other student in the class drops out, and all of the other groups of four have been functioning well up to that point.

6. Grace and Margaret are a pair of roommates in your class, and you have put them into different teams. The work that their two teams turn in is consistently nearly identical, including explanations and labeling of tables and graphs.

7. Nick's team turns in an assignment that is uncharacteristically poor. It was Nick's turn to write up the work, and this was the first paper for him to do so. When you hand the paper back, Nick is not present. His teammates are obviously shocked at their score. They approach you after class. They claim that they had left the meeting with all of the problems completed correctly including thought-out explanations, but they had left Nick to write up the final report on his own.

Diagnosis and Treatment of Sick Groups

Disease	Symptoms	Treatment
• Group not working together	• Reporter tells you • Student tells you • Different writing • Homework quizzes • Never converse • Scribe doesn't rotate	• Homework quizzes • Group evaluations • Group meeting in office hours • Discuss schedules — find meeting time • Point out that groups who work together are doing better
• Atmosphere in group not conducive to learning	• Reporter tells you • Student tells you	• Meeting in your office • Stress appropriate behavior to all group members
• Mismatched commitment to course	• Some people in group come to class, others don't • Same scribe • Differences in quizzes	• Discuss goals (personal and group) with each other • Reassure the students who are trying • Drop lowest scores
• Group works together but not everyone learns	• Homework quizzes • Quizzes in general • Exam scores do not match homework scores	• Meeting in your office • Emphasize the role of the clarifier • Encourage students to participate more actively in meetings

Hints on How to Make Groups Work[1]

Things that Help

1. Know each others' names.

2. Make sure that you begin by having one person read the problem.

3. Make sure everyone listens to everyone else.

4. Find a way for everyone to express varying opinions in a friendly way.

5. Be able to disagree without fighting, and be able to reach a resolution of a disagreement.

6. Make sure that *everyone* is equally involved.

7. Make sure that the group sets the correct pace for itself, without rushing or dragging.

8. Make sure consensus is reached.

Things to Avoid

1. Make sure cliques don't form.

2. Avoid leaving anyone out.

3. Don't "lay back" and hide from the group; participate!

4. Avoid making others feel "dumb."

5. Avoid talking all the time.

6. Avoid straying off the topic.

7. Avoid rushing to finish before everyone understands.

---

[1] By Bob Megginson's University of Michigan Math 105 Section, Fall 1996. Used with permission.

Team Evaluation Form

Name: _____
Date: _____

Please enter the names of your team members and evaluate their contributions as follows:

Not a real strength = 0
Okay = 1
A real strength = 2

Name of Team Member			
Attendance at Group Meetings			
Has read the material and tried to work the problems			
Comes to the meetings on time			
Helps to keep the group going			
Is willing to listen to others			
Puts effort into the process			
Is willing to disagree			
Knows whether other group members understand			
Helps assure that everyone understands the solution			
Describe the group member *			
Would you again be in a group with this person ? (Y/N)			

* Some examples (suggestions from the past) are: Worker, Leader, Leech, Obnoxious Type-A Personality, Sloth.

Please circle your response to each of the following questions:

Being part of this group helped me understand the material.
Strongly Agree   Agree   Neutral   Disagree   Strongly Disagree

Meeting with this group was better than working problems on my own.
Strongly Agree   Agree   Neutral   Disagree   Strongly Disagree

Meeting with this group was a good experience.
Strongly Agree   Agree   Neutral   Disagree   Strongly Disagree

What changes would you make to improve your group or group experience?
(*Please use other side.*)

7.7   Suggested Reading

- Barbara Reynolds, Nancy Hagelgans, Keith Schwingendorf, Draga Vidakovic, Ed Dubinsky, Mazen Shahin, and G. Joseph Wimbish, eds. *"A Practical Guide to Cooperative Learning in Collegiate Mathematics. MAA Notes #37."* [95]

  This book presents a theory of cooperative learning that revolves around groups that stay together for the course of an entire semester, and the discussion is relevant for either in-class groups or after-class groups.

  In Chapter 3, the authors discuss group formation. The issues they consider are the size of the group, the composition of the group according to several factors, the process of group formation, whether to involve students in the process of group formation, and the "lifetime" of a group.

  In Chapter 6, the authors describe a number of areas that students struggle with when they try to work in cooperative groups. The authors classify these as difficulties connected with (1) modes of operation, (2) organizational issues, (3) group dynamics, (4) individuals and (5) the rest of the world. As the breadth of some of these categories may suggest, some of the difficulties are simply too general to offer useful, concrete advice, and others may stem from deeper cultural issues that are very difficult to address in any event, and perhaps particularly so in the context of a math class. The authors' suggestions for dealing with difficulties primarily revolve around prevention by trying to convince the students of the efficacy of cooperative learning, and intervention by regular meetings between the instructor and the groups.

- Richard Chang. *"Success Through Teamwork."* [27]

  This is a book written for business managers and leaders, rather than for academics. The book is written in a very clear and straight forward fashion, and it identifies obstacles to teamwork that individuals may encounter. The book also includes a section on dealing with conflicts between team members, which is something that some students are very reluctant to try to resolve themselves. Instead, some students expect the class instructor to either take their side against the other team members, or to somehow resolve the situation. Because this book was written for managers who want to avoid "micromanaging" their employees, this book takes the viewpoint that most of what the team does will take place away from the immediate supervision of the manager. Thus, it presents a similar scenario to homework teams, where most of the teamwork takes place outside of the classroom.

- Ann Donellon. *"Team Talk: The Power of Language in Team Dynamics."* [41]

  This book is also written for business people, and is divided into two sections. In the first section, the author outlines a theory of the conflicting beliefs and attitudes that members of business teams must balance in order to effectively work together. This may be helpful in broadening an instructor's view of what else, besides getting to grips with the mathematics, students involved in homework teams must do to successfully complete their assignments together. The second part of the book describes case studies of four business teams. By analyzing the language that the team members use during their meetings, it attempts to describe why some teams successfully work together and why others do not.

- Ed Dubinsky, David Mathews and Barbara Reynolds, eds. *"Readings in Cooperative Learning for Undergraduate Mathematics. MAA Notes #44"* [43]

  This is a fairly large collection of articles describing various aspects of cooperative learning. Not all of the articles are principally concerned with mathematics—although they certainly have implications for mathematics—nor are all of the articles concerned with cooperative learning in college-level courses with college-age students. Perhaps the most pragmatically relevant of the articles are to be found in the "Implementation Issues" section of the book. Some of the articles here describe very specialized implementations of cooperative learning (for example, to computer laboratory settings), and it is up to the reader to determine what can be generalized from such accounts to other implementations of cooperative learning.

- Elizabeth C. Rogers, Barbara E. Reynolds, Neil A. Davidson, and Anthony D. Thomas, eds. *"Cooperative Learning in Undergraduate Mathematics: Issues, and Strategies that Work. MAA Notes #55."* [99]

Although mainly concerned with cooperative learning in class, much of the content of this book may be readily adapted to use with cooperative homework teams. For example, the creation of a positive social climate in class is one practice that may help teams that meet outside of classtime to function in a more harmonious fashion. Likewise, the discussion of theories of learning in group situations may inform the instructor's selection (or creation) of problems for team homework assignments. This book also includes strategies to address very practical issues, such as forming groups, which can be readily adapted to supplement the methods described here.

Chapter 8

Teaching During Office Hours

As all instructors are aware, teaching opportunities are not confined to the classroom. Office encounters provide one outlet for additional, and in this case more focused and personalized, instruction. Unfortunately, some instructors see office hours as an annoying burden and a waste of time that, in their eyes, could be more fruitfully spent on research or other duties. In addition, many students are reluctant to utilize office hours, because of various beliefs and perceptions that they hold.

Most instructors readily recognize that a primary function of one-on-one teaching in office hours is to provide students with individualized instruction and help. However, few may have reflected upon the other possibilities that such encounters afford—alleviation of mathematics anxiety, motivation, nurturing, retention, feedback (for the instructor as well as the student), etc. For this reason, we begin the meeting with a discussion of the purposes of holding office hours.

Students, on the other hand, often see office hours as a means to get "unstuck," especially in a mathematics class. Their primary motivation for coming to office hours will then be to enhance their performance on a current homework assignment, or an upcoming quiz or exam. Unfortunately, there are many motivational obstructions that hinder students from coming to office hours for help. Thus, we have included a section of the meeting on ways to encourage students to come to office hours.

Arguably the most important component of one-on-one teaching in a mathematics class is teaching problem solving. The final activity of this meeting asks the instructors to reflect upon what it means to teach problem solving. In particular, we stress encouraging student thinking, assessing student understanding, "helping" rather than "telling," and pointing out possible strategies. Because teaching problem solving is such a difficult and important part of mathematics instruction, a practice student encounter is included in the meeting. Reflection on the discussion in the meeting, as well as additional reading on problem solving is highly encouraged.

As with all topics covered in this training program, an overall theme of this section is the importance of professionalism. Instructors should take their duties during office hours seriously. Although keeping extra work around in case no students show up is an effective time management strategy, once students are present, the instructor should attempt to minimize distractions. Most importantly, the instructor should convey both verbally and non-verbally (both in and out of class) that students are welcome and encouraged to attend office hours all semester, not just right before an exam.

8.1   Description and Purpose of the Meeting

This meeting is intended to be a fifty minute session. Its purposes are to help instructors encourage student attendance at office hours, and to enable instructors to engage in meaningful one-on-one teaching during office hours. The meeting consists of a brainstorming session on ways to encourage student attendance at office hours, and facilitated practice on helping a student to learn to solve problems in one-on-one situations.

8.2 Goals for the Meeting

1. To provide instructors with a statement of institutional guidelines on office hours, if any exist.

2. To provide instructors an opportunity to think about the reasons for holding office hours, and to learn ways of encouraging students to attend office hours.

3. To provide instructors with an opportunity to practice the type of one-on-one teaching that occurs during office hours.

4. To provide instructors with a forum for discussing what makes a successful office hour encounter.

8.3 Preparation for the Meeting

1. Gather institutional or departmental guidelines on office hours, if any exist.

2. Choose a representative problem from your textbook that a student may bring to office hours.

3. Make copies of any additional readings that describe either teaching students to solve problems or "one-on-one" teaching.

4. Circulate the agenda for the meeting to all participants well in advance.

5. Make copies of meeting materials, and any questionnaires that you plan to distribute, and confirm attendance with participants.

8.4 Agenda for the Meeting (to be distributed to participants)

1. Discussion of the importance of providing quality teaching in a one-on-one setting.

2. Discussion of opinions of office hours based on your experience as undergraduate students.

3. Brainstorm ideas for getting students to come to office hours.

4. Discussion of effective one-on-one teaching in office hours, and a practice office hour encounter.

8.5 Outline for Running the Meeting

:01–:05 Introduce the meeting, and briefly review the agenda.

:05–:15 Lead a discussion on reasons for having office hours, and the important roles that office hours play in the success of the course and in the success of the department. In addition to the obvious benefit of the availability of "expert tutoring," office hours can perform a variety of functions in a mathematics course. Lead the group of instructors in making a list of these functions. The list may include

- to model expert problem solving,
- to detect and fill gaps in students' prior mathematics knowledge,
- to reduce mathematics anxiety by displaying a more intimate concern for the students' success,
- to teach study skills—such as how to assess an answer for correctness and reasonability,
- to motivate students who are not keeping up with the work,
- to encourage struggling students,
- to work with groups on group functioning,
- to teach calculator or computer skills and techniques,
- to encourage students' broader interest in mathematics,
- to give proficiency exams or make-up quizzes,

- to help the instructor stay attuned to the particular difficulties that students have with the material,

- to enable the instructor to get to know some students more personally and vice versa,

- to provide more focused opportunities for assessing student understanding and to provide more immediate feedback on student performance,

- to require the instructor to find alternative explanations for concepts—thereby improving the instructor's classroom skills, and

- to further facilitate a change in the students' perceptions of mathematics from a set of rules and procedures to a way of thinking.

:15–:25 Have the instructors fill out the sheet "Opinions of Office Hours" based on their experiences, both as students and as instructors. Lead a brief discussion on what motivates students to attend office hours, and what factors may discourage students from attending office hours, based on their responses to the handout. Some of the ideas may include

Reasons a student may go to office hours:

- to get help or hints on an assignment,

- to get clarification on a concept or idea,

- to be noticed by the instructor (convey eagerness),

- to get to know the instructor more personally,

- a lack of confidence in the student's ability to do the mathematics alone,

- to discuss the student's progress or course policies, or

- because the instructor encouraged (or required!) the students to attend.

Reasons a student may not go to office hours:

- mathematics anxiety,

- lack of approachability of instructor,

- gender or cultural issues may affect a student's willingness to seek help,

- laziness,

- office hours conflict with other classes or extracurricular activities,

- office hours are overly crowded with students, or

- (we hope!) the student is doing well and doesn't feel the need for personalized help.

Make a particular point to discuss the difference between the experiences and perceptions of the instructor and the experiences and perceptions of their students.

Pass out the sheet "Ideas for Getting Students to Come to Office Hours." Give the instructors a few minutes to read the ideas, and then facilitate a short discussion, where you collect new ideas from the instructors or discuss their reactions to the listed ideas.

Alternatively, you could refrain from passing out the ideas sheet, and instead have a short brainstorming session to generate a list of ideas within your group of instructors.

:25–:45 First, facilitate a five-minute discussion on what to do and what not to do when helping a student learn to solve a problem in office hours. If the group of instructors does not already do so, emphasize the importance of student understanding, particularly ways of assessing and enhancing it during the encounter. Some points may include

- get the student to think (instead of doing the thinking for them),

- concentrate more on the process and less on the "right answer,"

- get the student to reflect on what is being learned—especially how it is applied and how it relates to other material in the course,
- make the student do the writing (do not pick up a pencil),
- ask the student questions that probe for understanding,
- have the student work on a related but slightly different problem,
- give advice on problem solving strategies while refraining from solving the problem yourself, and
- encourage the student.

Count the instructors off into groups of three. In each group of three, assign one each of student, instructor, and observer for the upcoming role play. Give the "instructors" and "students" a copy of the exercise to look over for a couple minutes. Give the "observers" a copy of the handout "Office Hour Workshop Observer Checklist." Ask the groups to role play an office encounter, while the observer makes notes with the help of the checklist. Ask the "students" to try to approximate an actual student (i.e., be stuck, yet *reasonably* so). Allow about ten minutes for the role-play activity, including set-up time.

Finally, call the instructors' attention to you and facilitate a five-minute discussion based on the observers' responses to the checklists, and the impressions of those role-playing the encounter. In particular, discuss things that the "instructors" did that seemed particularly effective in assessing or enhancing the "students" understanding of the problem. Encourage the instructors to reflect upon effective problem solving strategies and ways to teach students *how to solve problems*, not merely how to do the problem that was brought to office hours. Give suggestions for further reading for those instructors who are interested in pursuing their knowledge further.

Some instructors may be quite resistant to the idea of participating in a role-play. Listed below are some strategies for getting reluctant instructors to participate.

- Acknowledge that the situation is artificial; nevertheless, it offers an opportunity to practice skills and receive feedback on them.
- Point out that role-playing in this way is an opportunity, not an obligation, but it is an opportunity that can only work if everyone participates.
- If someone has a serious problem being either the student or the instructor, have that individual switch assignments with someone who is to be an observer.

If you do not feel that your instructors will be willing to participate in a role-play, you could instead ask them to read the handout entitled "Office Hours Vignette." Ask them to use the checklist to assess the vignette, and then use their responses as a point of departure for the discussion. This approach has the definite advantage that it is more comfortable for the participants. We do feel that it does have disadvantages, however. For example, observing a role-play allows one to consider nonverbal communication, whereas reading a vignette does not. In addition, a role-play allows one person to practice one-on-one teaching, albeit in an artificial environment.

:45–:50 Wrap up the meeting with a word of thanks, and (as appropriate) time for participants' questions, a summary, suggestions for additional readings, or a questionnaire.

8.6   Meeting Materials

1. Copies of the handout entitled "Opinions of Office Hours."

2. Copies of the handout entitled "Ideas for Getting Students to Come to Office Hours."

3. Copies of the handout entitled "Office Hour Workshop Observer Checklist."

4. Copies of a problem from the departmental textbook that a student may bring to office hours for help, if you decide to do a role-play.

5. Copies of the handout entitled "Office Hours Vignette," if you decide to use it.

Opinions of Office Hours

Write down your answers to the following questions.

1. Think of your experience as a student. What are some reasons you have gone or would go to office hours? What are some reasons you would not go?

<u>GO</u>   <u>NOT GO</u>

2. Think of some of your typical students. What are some reasons that you believe they would go to office hours? What are some reasons they would not go?

<u>GO</u>   <u>NOT GO</u>

This activity adapted from one developed by the University of Michigan Center for Research on Learning and Teaching. Used with permission.

Ideas for Getting Students
to Come to Office Hours

1. Hold some or all office hours in a more attractive environment: coffee bar, departmental lounge, etc.

2. Announce that you know a particular aspect of an assignment may be hard, and that office hours on a particular day will be focused on that difficult portion.

3. Hold some office hours in a public computer site if any of the work of the course involves computers outside of class/lab/discussion.

4. In grading homework assignments, issue an invitation along with a correction: "The right answer is . . . . You will need this concept later in the course, so if it is not clear to you, please see me in office hours."

5. Remind students often at the end of the class session of when office hours are.

6. Offer special sessions to get students in the habit of coming in with problems, and to save yourself trouble (for example, how to use a special graphing calculator or other useful skills for the course).

7. Make sure to give each student at least one affirmative comment for coming to see you with their questions: "I'm glad you came to office hours instead of struggling even longer with this," or "I can tell this part is giving you trouble; I'm glad you came to see me with it."

8. When there are not other students waiting, take a moment to ask at the end of the session how the course is going for this student. It may give you good information, and it helps build a connection with the student.

9. Try to schedule office hours at different times (hours and days) so that most students can attend at least some of them, even if they have jobs and lots of other classes. Be willing to meet by appointment if a student truly cannot attend your regularly scheduled office hours.

10.

11.

This handout was developed by the University of Michigan Center for Research on Learning and Teaching. Used with permission.

Office Hour Workshop
Observer Checklist

Tally the number of times the Listener/Instructor does one of the following:

1. Prompts the problem solver to read or state the problem out loud. ____

2. Encourages the problem solver to think aloud.____

3. Asks for information, examples, explanations, or clarification.____

4. Offers information, examples, explanations, or clarification.____

5. Offers problem solving advice.____

6. Paraphrases or summarizes what the problem solver has said.____

7. Gives reassurance or praise.____

Who actually solved the problem, the problem solver or the listener?

Any suggestions for the listener? (Note down anything the listener said that you think was particularly effective or that you think needs to be discussed.)

This handout was developed by the University of Michigan Center for Research on Learning and Teaching. Used with permission.

Office Hour Vignette
Transcript of Instructor and Student

Student: I couldn't get #21.

Instructor: You couldn't get 21. Let's see. Why don't you read the problem to me?

S: At a sand and gravel plant, sand is falling off a conveyer and onto a conical pile at a rate of 10 cubic feet per minute. The diameter of the base of the cone is approximately three times the altitude. At what rate is the height of the pile changing when the pile is 15 feet high?[1]

I: OK. How far did you get?

S: Uh. Well. I got that $dV/dt = 10$.

I: Good. Write that down. Now, what is V?

S: Volume.

I: How do you know?

S: Um ... the units are cubic feet.

I: Right. The units of the rate are cubic feet per minute, so you know that this is the rate of change of the volume. What else did you do?

S: I found the formula for a cone.

I: You know the formula for the volume of a cone? Write that down.

S: OK. (Writes $V = \frac{\pi r^2 h}{3}$.)

I: What are h and r?

S: Uh, h is height, and r is radius.

I: How does that help?

S: It has volume in it, and I know that the height of the pile is 15 feet.

I: OK. But these are related rates problems. We know the rate of change of volume. Do we know any other rates, or are we asked for any other rates?

S: Um ... (reads) ... we're asked for the rate of the height.

I: Good. How do you write that in symbols?

S: It's dh/dt. (Writes $dh/dt =$?)

I: Good. Are there any other rates?

S: I don't think so.

I: OK. Hmm. Is that going to be a problem? I mean, do we need to know anything else to finish the problem?

S: Oh, right! Here's where I got stuck. Um. I didn't know what to do with the r.

I: Why is the r a problem?

S: I don't have a rate for it.

---

[1]Problem #21 on page 146 of Larson, Hostetler, and Edwards [73].

I: That's right. When you take the derivative with respect to t, a dr/dt will pop up. Can you relate r to any of the other variables? I mean, can you get rid of it somehow, before you take derivatives?

S: The diameter is three times the altitude. Diameter is kinda like radius. Is the altitude the height?

I: Yes. That's right.

S: So $D = 3h$. Oh, I see. So $2r = 3h$ (writing as he goes). So $r = \frac{3}{2}h$.

I: Now, what can you do with that?

S: Then my formula for V is (writes $V = \frac{\pi(\frac{3}{2}h)^2 h}{3}$). So I take the derivative? Ugh!

I: Well. You may want to simplify first.

S: Duh. (simplifies) Then, I take the derivative, and I plug in, um, plug in $dV/dt = 10$ and, uh... (reads).., $h = 15$. Then I can solve for dh/dt. (working it all out) Yeah, all right.

I: Don't forget units!

S: Units. Height is feet, so it's feet per minute.

I: Good job. Any others?

8.7   Suggested Reading

- Leo Lambert, Stacey Lane Tice, and Patricia Featherstone, eds. *"University Teaching. A Guide for Graduate Students."* [71]

  Chapter 6 describes some of the practicalities of office hours (e.g., announce them), and emphasizes the need to listen to students to try to determine the source of the students' difficulty, rather than simply repeating previous presentations of the material.

- Steven Krantz. *"How to Teach Mathematics."* [69]

  Chapter 2, Section 6 describes office hours. This section emphasizes the opportunities that office hours can present for the instructor to get to know students, and to better get to know students as learners of mathematics.

- A. Whimbey and J. Lockhead. *"Problem Solving and Comprehension."* [120]

  The authors catalogue and classify sources and types of errors that a student may make when trying to solve problems. These can be used as a framework for assessing students problem solving techniques in order to help them become better problem solvers.

- George Pólya. *"How to Solve It."* [89]

  This classic provides a four-step procedure for solving mathematical problems–understand the problem, devise a plan, carry out the plan, and look back. Pólya gives advice to teachers aimed at helping students learn to solve problems. He suggests how to help students unobtrusively by asking questions or indicating steps that could have occurred to the student. He provides a dictionary of explicit heuristic suggestions and questions, such as "What is the unknown?" "What are the data?" and "Do you know a related problem?"

Chapter 9

Establishing and Maintaining Control in Your Classroom

The introduction to this chapter is slightly longer than is typical. The idea of this introduction is to provide instructor trainers with a basic idea of what issues may be described under the title of "Control in the Classroom."

Instructor trainers are usually naturally talented and effective instructors who have a clear vision for what they want to make happen for the students in their class. Likewise, instructor trainers will often have a natural talent for clear communication, and will instinctively take steps to ensure that their classroom is an ideal environment for this communication to take place. As such, instructor trainers may have had little direct, personal experience with classrooms that are out of control. This was certainly the authors' experience when visiting the classrooms of some novice instructors for the first time. We found that in order to help instructors, we needed to reflect on exactly what we understood by "control" in the interactive classroom. The introduction to this chapter presents some ideas of what "control" in a mathematics classroom can mean, and areas where difficulties can arise.

The word "control" evokes a variety of responses from mathematics instructors. Some instructors respond that control is essential—without control over the classroom, students will create so many distractions that no learning is possible. The following short account of such a classroom (related by Steven Krantz [69]) vividly illustrates this possibility.

> I have seen large mathematics classes (of about 400 or more students) which looked like a cross between a rock concert and a Hieronymous Bosch painting. Private conversations and mini-dramas were taking place all over the room while groups of students roamed the aisles. The professor stood at the front of the room, bellowing away on his microphone, while a small percentage of students attempted to learn something.

Other instructors believe that students must feel free to explore mathematical concepts and to construct their own understandings of the mathematics. A classroom guided by such beliefs is described in an account by Jean MacGregor [43].

> Having read a chapter on the origins of life in their introductory textbook, students arrive at class and pick up worksheets. They gather around the teacher, who, using the phospholipids in egg yolk to mix oil and water, is demonstrating hydrophobic and hydrophilic characteristics of phospholipid molecules. Then dividing in groups of three or four, the students start on the worksheet problems. They diagram possible arrangements of phospholipids in the "primordial soup" that might have led to the first cell membranes, and then speculate on the sources of these molecules.

These instructors worry that control from the instructor may stifle the students' activity, or encourage students to substitute recitation of the instructor's words for the expression of their own ideas.

To understand what it means to "control the interactive mathematics classroom," we need to understand the nature of the interactive classroom, and in particular, the nature of the difficulties that may arise.

115

Most of the "Instructor Guides" included at the end of [23] offer some suggestions on what kinds of problems may occur, and give some suggestions of measures that instructors may take (the one from MIT is called "Problems with Students and Students with Problems," for example). These advice sets usually offer descriptions of and suggestions for difficult situations that instructors have encountered. The advice sets are usually indexed by specific student behavior (such as "overdependence on instructor," "aggressive, grade grubbing," "attracted to instructor," etc.) and are often principally concerned with problems that arise from close, interpersonal encounters between the instructor and individual students.

We suggest four categories of difficulties that instructors who are committed to creating and running an interactive mathematics classroom may encounter. These categories are motivated by our experience as instructors, and by discussions of difficulties in the teaching literature (see, for example [38], [48], [49], [51], [53], and [95]).

- **Implicit, mismatched ideas of course expectations between instructor and students.**

  This may include some students' assumptions that no explicit attendance policy means that absences from class do not need to be excused. This may also include the frustration that some instructors express when students do not attend class. This may include students' complaints that their instructor is a "tough grader." This may also include the occasional instructor comment that students "do not take enough time with their work."

- **Difficulties that arise from the nature of an interactive classroom.**

  These may be viewed as students' responses to the opportunities that the interactive classroom offers them to affect the pace of the class, influence the direction of the class, or to respond to the opportunities for interaction that the instructor creates in the classroom. What distinguishes these student responses from the ones that the interactive classroom is designed to foster, is that these student responses will not be directed towards a productive engagement with the subject matter of the class. Related to this is the phenomenon of students who have enjoyed a great deal of success (at least as measured by grades) in other instructional settings (such as so-called "Direct Instruction"). These students may be suspicious of (or even hostile to) the notion of an interactive classroom, [15], [16] and [101].

- **Difficulties between individual students and the administrative structure of the course.**

  For example, some universities have strict policies governing absences from exams, and make-up exams. Students sometimes ask an instructor to take an exam early or late for reasons other than those mentioned by official policy.

- **Highly interpersonal situations that arise between individual students and the instructor.**

  These include many of the "Problems with Students" that have been described in publications such as [23]. Mostly, these are situations that occur away from the classroom, although they will often have obvious repercussions for the classroom atmosphere. Examples include belligerent or aggressive responses to tests or grades.

We note that it is somewhat simplistic to suggest that every difficulty that an instructor encounters will fit neatly into this classification scheme. For example, a student who persistently starts off-topic conversations in the classroom may be thought of as responding to the interactive classroom. If the instructor believes that class time is too valuable to spend on idle "chit-chat," then this situation also has aspects of a mismatch of ideas or expectations for classroom behavior between the student and the instructor. We hope that these four categories will serve as some kind of organizational scheme to help instructor trainers to make sense of control issues, rather than a definite guide to every possible classroom scenario.

The purpose of this meeting is to discuss difficulties that instructors have encountered that are primarily difficulties arising from the nature of the interactive classroom. The question of dealing with the final category of difficulties—highly interpersonal situations that arise between individual students and the instructor—is dealt with in Chapter 14.

9.1 Description and Purpose of the Meeting

In preparing this meeting, we assume that the instructors have been encouraged to formulate appropriate learning goals for each class meeting, and that they are able to devise suitable plans for the class meeting that can potentially help students to achieve these learning goals. In particular, we assume that the difficulties in the classroom are not the result of the instructor trying to do too much—or too little—with the class time available.

The purpose of this meeting is two-fold. First, the meeting is a forum for instructors to think about and respond to the questions:

- What about the nature of an interactive classroom — and particularly the opportunities that it grants students to affect the conduct of the class — can stand in the way of achieving learning goals?

- What can instructors do so that the opportunities for student participation are undiminished, and yet the instructor is able to ensure that all learning goals will be adequately addressed?

Second, this meeting can serve as a support session for instructors facing classroom management issues. In this capacity, the meeting can give concrete support to instructors who feel that they cannot "control" their classes, and to encourage instructors to discuss the teaching challenges that they face with each other.

Overall, the meeting aims to both provide instructors with some kind of framework for analyzing the challenges they face, and to provide hope to instructors who are discouraged by control difficulties.

9.2 Goals for the Meeting

Some possible goals that this meeting may have are

1. To provide participants with an analytical framework for categorizing, and perhaps beginning to understand and put into perspective, the difficulties that they may have experienced in their classroom.

2. To have participants share classroom experiences that make them feel uncomfortable. Many meeting participants may not normally be able to speak about these issues with colleagues.

3. To have participants realize that others also face serious issues of control in their classes, and that they are not alone in their experience.

4. To have instructors think about the challenges that they and their colleagues face, and find workable solutions.

5. To alert instructors who have not experienced serious issues of control and power that such problems can occur, and what the nature of the problems can be.

6. To give participants a chance to reflect on their experiences, and to give them an opportunity to think about what they could do differently next time.

9.3 Preparation for the Meeting

Perhaps the most successful way to design this meeting is to build it around the experiences and needs of the participants. To do this, the meeting leader needs to be familiar both with the (1) kinds of issues that participants are facing in their classes—and how they have responded to these issues—and (2) what about the nature of an interactive classroom can contribute to the difficult situations that instructors face.

1. The meeting leader needs to find out what kinds of control issues the participants have been dealing with in their classes through interviews or a survey. If the professional development program includes classroom visits, then the visits and the subsequent interviews with instructors can provide much of this information.

2. If the meeting is to be one where the leader candidly expects people to share their experiences of losing control of their class, or of being challenged by students and not knowing what to do, then the leader may have to do some tone-setting work before hand. One way to do this is to write up a sheet describing some difficult situations, and distribute this to participants (perhaps prior to the meeting). The idea is to construct descriptions of situations in which participants will recognize something of their own experience, but which are not so pointed that participants will feel vulnerable and overexposed. A sample set of difficult situations is appended as the handout, "Classroom Situations."

3. If it seems appropriate, the meeting leader could invite experienced instructors who have dealt with control issues in their own classes to attend the meeting. Many experienced instructors are adept at setting the tone of their classes so that control is not a problem from the outset. It is probably helpful for meeting participants to have access to people with these kinds of skills, but it is important for participants not to dwell on what they think they have done wrong. It is particularly important for meeting participants not to get the idea that they have already failed in their classes, because they did not set them up on the same terms as an experienced instructor may have. It is also important for meeting participants not to get the idea that their situations are hopeless. With this in mind, it may be profitable to ask experienced instructors who have recovered from problems with control and power. Their experiences of winning a class back will, at least in the short term, be more pragmatically useful to meeting participants.

4. The meeting leader may research some of the literature dealing with assertiveness, classroom behavior and teaching strategies, in order to synthesize a handout that relates some strategies for dealing with behavioral problems in general, or specific issues (for example, students talking out of turn). The following references (see the suggested readings at the end of this chapter and the bibliography for more) are a useful place to start.

    (a) McKeachie, Wilbert J. *Teaching Tips. 9th Edition.* Heath. [81]

    (b) Cangelosi, James S. *Classroom Management Strategies. 3rd Edition.* Longman. [21]

    (c) Rinne, Carl H. *Excellent Classroom Management.* Wadsworth. [97]

    The references [38], [48], [49], and [51] may also be valuable background reading.

5. Plan to end the meeting on a positive note. It may be helpful to consider exactly how to do this as part of the preparation for the meeting. For example, the meeting leader may make a summary of some of the things that are going well. For an alternative idea for wrapping up, see the outline for running the meeting.

6. If you plan to enforce any ground rules at the meeting, then plan them beforehand. Ground rules may be circulated to meeting participants before the meeting starts, or else they can be handed out and described at the beginning of the meeting.

7. Make sufficient copies of meeting materials that you think may be helpful to the participants. A handout entitled "Control in the Classroom" is appended to this meeting outline, which can serve as a starting point.

9.4   Agenda for the Meeting (to be distributed to participants)

Since this meeting will be driven by the needs of the participants, rather than by a set of skills that the meeting leader wishes to impart to the participants, it makes less sense to plan and circulate a very rigid agenda. Below is an example of a flexible agenda that could be circulated.

Example Agenda

Item 1. A discussion of problems that can arise with management of the interactive classroom. This will consist of a review of what research literature has to say about the nature of classroom difficulties, and what can be done about them. Our focus will be on problems that can arise from the nature of the interactive classroom—"off-topic" conversations, lack of connection between group work and other activities in the class, having the class slowed down too much, etc.—as opposed to problems that have more to do with individual students.

Item 2. Discussion with others of how you see your classroom situation, and what you feel you would like to see happen differently. Everybody present will have a chance to talk about the particular issues that they are dealing with, and to receive support, ideas and advice from other participants.

Item 3. What can be done now? What possibilities does the current situation present? There will be opportunities for participants to say what they would really like to change about their classroom situation, and receive ideas and advice from other participants, and experienced instructors to try to sort the situation out.

Item 4. Discussion of some classroom management strategies that people have tried in the past, and the results that they have seen.

Item 5. What can we learn from this? How can we conduct ourselves so that we don't have to go through this kind of an experience again and again and again?

9.5 Outline for Running the Meeting

As this meeting will be driven by the interests and thoughts of the instructors, rather than by a set of skills and an agenda, it is more appropriate to loosely plan some phases for the meeting than to follow a rigid schedule. Suggested phases, in approximate chronological order, are given below.

- Set ground rules, if ground rules are explicitly mentioned and used.

- Lead a tone-setting exercise, such as the "Student Situations" sheet. The meeting leader could ask participants to look over the list, and see if any of the situations described resonate with their own experiences. If so, ask them to discuss the situation with a neighbor. When conversation begins to flow more or less easily, the tone is probably appropriate for a wider, but still candid, discussion of the issues that individual participants face.

- Give a short "mini-lecture" on typical problems that seem to arise from interactive classrooms, and list ideas for what can be done to address these problems. Emphasize from the outset that the problems that you will concentrate on are problems that are mostly due to the nature of an interactive classroom, and the opportunities that it presents students, rather than problems with individuals in the class.

- Invite participants to share their experiences, and for other participants to support them. The sincere use of active listening and sensitive facilitation can be used here to draw out participants' experiences.

- Formulate concrete strategies and ideas for individual participants to try out when they go back to their classes. Pass out copies of the handout "Control in the Classroom" or a similar handout developed from a review of the literature.

- Wrap up the meeting on an up-beat note, trying to sincerely emphasize that there are positive things going on in the participants' classrooms. One way to do this is to emphasize the *solutions* and *ideas*, rather than the *problems*. Another way to end the meeting on a positive note is to make an analogy with "a black dot on white paper." Just as the eye is drawn to a black dot on a sheet of paper, ignoring the large expanse of white around the dot, the participant's self-criticism is drawn to the things in their class that are not going well. They lose sight of the many things that they *are* doing very well. If the meeting leader is well acquainted with each participant's teaching situation, and aware of many things that are going well, then the meeting could be wrapped up by reminding participants of the things that are working in their classes. The meeting leader should appreciate the need for sincerity. Some of the meeting participants may be desperately worried about how they appear, and anything less than total sincerity may be seen as a cheap ploy, and end up doing more harm than good.

- Have the participants evaluate the workshop.

9.6 Idea for Expediting the Meeting

Instead of holding a meeting that all instructors are required or encouraged to attend, circulate an invitation, asking people to attend if they feel that they would like to. The tone should be inviting, and very deliberately non-judgmental. That is, participants should not feel that they are admitting some kind of defeat by attending the meeting.

Colleagues,

In response to inquiries from interested instructors, we have planned a meeting to be held at <time> on <day and date>. The objectives of this get-together will be to discuss problems of classroom management, and what can be done to remedy problems. We hope that this will be an informal session where we are free to discuss not only some of what the literature has to say about classroom management, but also to listen to each others' classroom experiences.

If you have ideas, or if, like me, you would like to talk about experiences in your class in an informal and relaxed atmosphere, please consider yourself to be very warmly invited to attend.

Sincerely,

<meeting leader>.

9.7 Meeting Materials

1. Copies of the handout entitled "Control in the Classroom."

2. Handouts developed from the literature on Classroom Management.

3. References on Classroom Management.

4. Copies of the handout entitled "Classroom Control Situations."

Control in the Classroom

- When starting a group exercise, count the students off and have them physically move to a designated area of the classroom to meet the other members of their team and work on the assigned exercise. When planning or setting up a group activity, bear in mind the following.

  - If students work in their homework groups during class time, then they may work on their homework assignments, rather than the learning activities assigned.

  - Groups may work at different paces. If it appears that some groups consistently finish their learning activities well in advance of the rest of the class, plan some additional exercises for the very quick students to go on with.

- During a group exercise, closely monitor students' activity. Be on the look-out for students who seem to be talking about things other than the assigned learning activity. If it becomes clear that students are not working on the assigned learning activity, you may consider the following suggestions.

  - Ask the people who are talking where they are in the problem you have assigned.

  - If the students have finished the assigned activity, ask them to explain their work to the other people in their group or to write their solution up on the board.

- When ending a group exercise go to the front of the room, and loudly ask for everybody's attention. Repeat the request until the talking stops.

- When a student answers a question, and others are talking, thank the student for their contribution and say to the class, "<student's name> just made a really good point, and I think that you all should hear it." When the talking stops, ask the student to repeat their answer, and thank them again. If the talking does not stop in a reasonable period of time, consider the following.

  - Tell the class that grasping the next point is absolutely critical (or write something on the board to this effect).

  - Make an obvious point of waiting for the talking to completely stop before making the point.

  - Try the "speak–write–speak" strategy. That is, state the point, write a note summarizing the point on the board and finally state the point again.

  - Directly ask the student who refuses to stop talking (e.g., address the student by name) a question about the point that has just been made.

We note that these suggestions have some measure of risk associated with them. For example, if the instructor makes a point of appearing to wait impatiently until the talking subsides and yet the talking continues unabated, then the instructor may appear foolish.

- Some instructors are able to employ silence as a means to maintain order. We note that this strategy may be more appropriate to instructor-centered portions of the lesson, where the sound of the instructor's voice is the norm, and a (unusual) period of silence is a clue to students that something is happening. While the students are actively working on learning mathematics, instructor silences are less likely to be an effective means of gaining students' attention.

Classroom Control Situations[1]

- Students engage in activities not related to the subject matter being studied.

Example

To follow up on a short lecture on concavity, the instructor asked the students to break into groups and try a problem from the text. The instructor repeated the problem and page numbers several times, and wrote them on the board. Most of the students were sitting next to their friends or members of their homework groups, and this is how they formed groups. Many of these groups began long and loud conversations discussing problems from the homework assignment due later that week. In the meantime, the instructor walked around from group to group. Despite the presence of the instructor, many groups remained focused on their homework.

What can be done?

- Set ground rules—and stick to them!—for what kinds of activity is acceptable during class.

- At the beginning of a group activity, take a minute to explain the goals of the activity to the students, with emphasis on what they will know or be able to do better as a result of participating.

- Occasionally ask all of the groups to go and work the activity on the board.

- If you spot a student or a group who is not working on the activity you have assigned, ask them how much they have completed. If they reply, "Not much," then suggest that they should get to work on the assigned activity. Repeat (with variations) as necessary.

- Students do not interact with each other when appropriate to do so.

Example

The instructor has prepared an activity on a worksheet. The activity provides the students with a set of data giving information about the number of animals and the number of tourists on a coral reef. The idea of the activity is to have the students calculate a regression line to relate the number of animals to the number of tourists. The instructor distributes the worksheets to the students, and tells them to work on it in pairs.

As soon as they receive the worksheet, most of the students read it and begin to write in the spaces provided. The only sound in the room is the instructor's voice saying, "It's okay to work on this together, guys."

What can be done?

- Form the students into groups yourself at the beginning of the activity, and if the students are new to group work, ask the students to begin by introducing themselves.

- Distribute only one copy of the worksheet to each group, and appoint someone in the group to read it to the others at the beginning of the activity.

- Try to arrive in the classroom a little before the students and, if possible, rearrange all of the chairs so that students will be facing each other when they take their seats.

- The difficulties that some students seem to have understanding the work at hand slow the pace of class to the point where it is impossible to finish.

---

[1] Adapted from the article by DeLong and Winter, [38].

Example

The instructor has prepared a pair of examples to demonstrate the relationship between the concavity of a graph and the average rate of change of the graph. The lecture is interactive in that the instructor asks students to calculate the average rates of change on their calculators. When the students report the results, the instructor records these in a table on the board.

"Wait a minute," says one student, "Where are these numbers coming from?" The instructor demonstrates one of the calculations in detail on the board while the other students look on.

The instructor then asks the class to look at the patterns in the numbers on the board and make a conjecture. One student puts up her hand and comments, "When the graph is concave down, the average rate of change is decreasing." The instructor agrees, and superimposes secant lines on a concave down graph to illustrate this point.

"Wait a minute," says the student from before. "What's this supposed to be showing now?" While the other students in the class look on, the instructor explains that as you go from left to right the slope of the secant lines decreases.

What can be done?

- Remember that this kind of situation is often more frustrating for the student who doesn't "get it" than for the students who do.

- Remember that the student is usually trying to do exactly what you want him or her to do– understand the mathematics—it is simply the circumstances that are inconvenient.

- Another possibility to bear in mind is that some, especially first year, students are accustomed to a slower pace in class, and may expect to be able to pick everything up during class time. Remind the students that they should also be spending quite a lot of time outside of class working to understand the material. Some prescriptions suggest two hours outside class for every hour in class.

- Suggest to the student that you meet after class, or during office hours, to discuss the point in detail. For now, the student should try to develop some idea of what's going on, and you'll help them to develop a complete picture later on.

• Some students dominate the proceedings.

Example

Wrapping up an activity on linear regression, the instructor calls for attention. Waiting until the students stop talking, the instructor asks the class what was the equation of the line they calculated. Harry calls out his answer. The instructor writes this on the board and asks, "Okay, what does a slope of -0.13 mean here?"

Again, Harry speaks up, "For every extra tourist, there are 0.13 less animals per square foot." The instructor nods, and asks, "What about the intercepts, what do they mean?" Again, Harry responds, "The 27.75 means that there will be this many animals when there are no tourists, and the 213.46 means that this many tourists will kill all of the animals."

The instructor nods and asks, "Is there a causal relationship here?" Harry doesn't hesitate, "Well, yeah. It seems like the tourists are killing off the wildlife."

What can be done?

- Direct some of the questions to particular students, as well as leaving some questions open to volunteers.

Some instructors—and students—object to this, because they feel that students may be made uncomfortable by questions that they are not immediately able to answer. To ameliorate this concern, try to:

* Make the classroom an environment where genuine attempts to get to grips with and think about problems are valued.
* Make an effort to call on all students equally.
* Call on students in a pattern. For example, every other seat.
* If students do have strong fears, they will sometimes approach you and ask you not to call on them. Respect the particular student's wishes.

9.8 Suggested Reading

- Donald Finkel and G. Stephen Monk. "Teachers and Learning Groups: Dissolution of the Atlas Complex." [51]

  This article centers around a concept that the authors call "The Atlas Complex." This can be characterized as the tendency of instructors and students to assume that the instructor is ultimately responsible for everything that goes on in a classroom. In particular, the instructor is ultimately responsible for whether or not the students learn. One of the unfortunate corollaries that has particular relevance for mathematics is that instructors cannot expect students to exhibit anything but a poor rendition of the intellectual performance they witness in the classroom. In particular, instructors cannot expect students to be able to learn or work out anything that they have not explicitly "seen." Finkel and Monk suggest that student-centered classrooms may provide a way out of this situation. They describe problems that faculty who attempt to "dissolve the Atlas complex" may encounter.

- Steven Krantz. *"How to Teach Mathematics."* [69]

  This book contains useful advice to beginning instructors. Because the book concentrates on the "lecture method" of instruction, not all of the advice is relevant to the interactive classroom. On some subjects (such as dealing with students who persistently talk out of turn), the author is very explicit, even suggesting sentences that instructors may say. However, although the author points out that sometimes approaches other than the lecture are effective (for example, he suggests some discussion if the class is small enough), he is not very clear on how to start and facilitate a productive discussion. This information would be of particular relevance to mathematicians, because the nature of mathematics is perceived to be very different from "discussible" subjects, such as literature or politics. Mathematicians could well benefit from ideas on how to formulate mathematical problems in ways that make students think that there is actually something to discuss, rather than just a "right answer" to sit and wait for.

- James Cangelosi. *"Classroom Management Strategies. Third Edition."* [21]

  Although written for pre-service and in-service elementary, middle and high school teachers, this book describes features of a classroom environment that may be useful for college instructors. The author's main point is that there is a positive correlation between student achievement (in all subjects) and the amount of time spent actively working in class ("time on task"). The author also notes that there is no demonstrable, positive effect of creating a "warm and fuzzy" classroom atmosphere (at least in terms of student achievement). Instead a "businesslike atmosphere" seems to be the most effective at encouraging students to spend more time working on learning activities. The author provides illustrations of the kind of classroom environment he envisions, and suggestions for how the "businesslike atmosphere" can be created.

- Matthew DeLong and Dale Winter. "Addressing Difficulties with Student-Centered Instruction." [38]

  This article is not specifically about problems that arise from the nature of the interactive mathematics classroom. Instead, the authors concentrate on the related topic of problems with instructors' implementations of student-centered instruction in mathematics. The authors identify four areas that beginning instructors seem to struggle with—the use of questions, management of cooperative learning, management of instructor-centered activities, and giving up forms of control—and provide a wealth of suggestions for instructors trying to deal with these kinds of implementation problems.

Chapter 10

Proctoring Tests and Examinations

One of the "nuts and bolts" issues that all teachers must get to grips with is proctoring students during tests and exams, and taking appropriate action if they observe students cheating. Much of the advice and many of the precepts set out in this meeting will be obvious to experienced and capable instructors. In our experience, many beginning instructors are aware of the possibility of cheating and want to prevent it. However, most do not have clear ideas of exactly how to go about this. We have worked with instructors who sit at the front of the exam room, obviously engrossed in a book, and subsequently complain about the high number of "suspicious similarities" between exam papers. Occasionally, we have run into instructors who seem to be a little too zealous in their desire, and too severe in their methods, to eliminate cheating. We have found it important to try to be very clear about the limits of the proctor's authority. Finally, most universities have set procedures and official channels for dealing with cases of suspected cheating. Instructors who are new to teaching or new to a mathematics department need to be informed of these processes, and need to be made aware of the appropriate contacts within the department, should a situation arise.

10.1 Description and Purpose of the Meeting

This meeting is intended to be a short session (thirty minutes would probably be adequate). The meeting is intended to provide a clear statement of what is expected of instructors regarding examinations and what to do if they suspect that students are cheating. The meeting can also be used to give instructors some ideas for returning papers to students, and the kinds of questions that arise regarding cheating. It is not our intention to describe the definitive guide to dealing with cheating in all its many forms during this meeting. This meeting would, ideally, be held just in advance of the first major examination in the course. There should be opportunities for instructors to have their questions answered, but this meeting should otherwise be short, businesslike and informative.

10.2 Goals for the Meeting

1. To provide instructors with a statement of institutional guidelines regarding examinations and cheating.

2. To provide instructors with a clear idea of what they are expected to do when they give and proctor examinations.

3. To provide instructors with a clear idea of what they should do if they suspect that students are attempting to cheat on the exam.

4. To provide instructors with ideas for how and when to return graded exam papers.

5. To provide instructors with a forum to have their questions about the examination process answered.

10.3   Preparation for the Meeting

1. Gather institutional or departmental guidelines on examinations and cheating, if any such guidelines exist.

2. Review and modify the handout on examinations and proctoring to bring it into line with the standards and customs of your institution.

3. Circulate the agenda for the meeting to all participants well in advance.

4. Make copies of meeting materials, and any questionnaires that you plan to distribute, and confirm attendance with participants.

10.4   Agenda for the Meeting (to be distributed to participants)

1. Outline of departmental and institutional regulations examinations and cheating.

2. Statement of instructor responsibilities for the examinations.

3. Discussion of what to do if you suspect cheating in the exam room.

4. Discussion of different methods for returning graded papers to the class, and steps that you can take to try and prevent alteration of papers.

10.5   Outline for Running the Meeting

:01–:05 Introduce the meeting, and briefly review the agenda.

:05–:10 Distribute the copies of the modified handout. Focus participants' attention on the portions of the handout dealing with giving and proctoring the examination. Give the participants a chance to read over these sections, and then highlight the critical points. Reserve a little time for questions.

:10–:20 Focus participants' attention on the sections of the handout on cheating. Give them an opportunity to read these sections, and then highlight the critical points. Make sure that the standards and conventions of your institution are clearly understood by all participants. It can also be helpful to ask experienced instructors who are attending the meeting to relate incidents where they have suspected cheating. Ask them to describe what they saw, and what action they took. Reserve some time for questions.

:20–:28 Focus participants' attention on the sections of the handout on returning papers to students. Give them an opportunity to read these sections, and then highlight the critical points. Make sure that the standards and conventions of your institution are clearly understood by all participants. It can also be helpful to ask experienced instructors who are attending the meeting to relate their own experiences of returning papers to students. Ask them to describe what they did, and if there were any problems or difficulties. If there were problems, ask the experienced instructors what they plan to do differently this time to try and avoid those same problems. Reserve some time for questions.

:28–:30 Wrap up the meeting with a word of thanks, and (as appropriate) time for participants' questions, a summary, suggestions for additional readings, or a questionnaire.

10.6   Meeting Materials

1. Copies of institutional or departmental guidelines or regulations covering examinations, proctoring and cheating.

2. Copies of the handout entitled "Proctoring Tests and Examinations."

Proctoring Tests and Examinations

1. Cheating

Cheating is a difficult, personally demanding and unpleasant problem to deal with. On uniform exams, worrying behaviors include:

- Looking at other peoples' work.
- Showing work to other students, or consciously allowing other people to see work.
- Using materials or references not permitted on the exam (inside and outside the exam room).

If you observe behavior that you consider to be cheating, and you share the exam room with another instructor, please tell them what you have seen. Remember that your institution will have established procedures for dealing with academic dishonesty. These procedures exist both to protect students and to protect you. Don't be tempted to apply "home-made" justice. It may pay to make a note of who is sitting nearby the offending student(s), so that papers can be compared later.

Remember that many innocent behaviors (like looking skyward as you promise to be good from now on if only you can get the answer to #3 ...) may look like cheating. Please be absolutely certain that cheating has or is occurring before taking any action.

Remember that when dealing with cheating, an ounce of prevention is worth (at least) a pound of cure. Perhaps the best way of preventing cheating is through good proctoring.

2. Proctoring

While students are taking the exam, it is important for you to proctor actively. Please do not bring reading material or other kinds of amusements to the exam room. By sitting at the front of the room, and not paying attention to the people taking the test, you are encouraging dishonest behavior.

As people take the test, proctors should move around the room, keeping tabs on what is happening. Try not to be too obtrusive, but make students feel that they are being observed—that you will see if they try to do something blatantly dishonest. One trick is to make (and briefly sustain) eye contact with any person whose behavior is making you nervous or suspicious.

You may find it helpful to make a list of the students from your class who are in the exam room. Although it is rare, occasionally a student will fail to appear for the exam. With a definitive list of people who were present in the exam room, it is easy to be completely sure of whether the student was there or not (especially if the student's recollection of events differs from your own). At the end of the exam, this list can be used to check off exams as you receive them. By doing this, you can be absolutely sure that everybody has handed in their exam. Before leaving the exam room, make a thorough search of the room, including waste-paper baskets, under desks and chairs. Again, it is rare, but not completely unknown for a desperate student to fail to hand their exam in, and later claim that the instructor lost it in the hopes of being able to take the exam a second time.

Sometimes, during the course of the exam, a student will want to leave the exam early. This is usually perfectly acceptable. If several classes take an exam in the same room, things can become very confused. For this reason, if a student who is not in your class wants to hand you their paper, please either pass the paper immediately to the appropriate instructor, or else simply refuse to accept the paper, and insist that the student hand it to their own instructor. If the appropriate instructor is indisposed for some reason, have the student wait until their instructor returns.

During the course of an exam, it is normal for several students to have need of the bathroom. It is not common (but then again, also not unheard of) for students to be escorted to and from the bathroom, but it is helpful if you know where the closest bathroom is. If a student is gone for a long time, or asks to leave the room repeatedly, then it is probably time to get suspicious, and perhaps time to insist on an escort.

3. **Handing Back Papers**

When returning papers to students, it is important for students to understand what mistakes they made. Many students feel that it is important to understand why they weren't given more points on the exam.

It is not clear that these two issues are complimentary; spending an entire class period trying to bring out the meaning of the exam questions can be a futile effort if you are constantly being interrupted by protests, "But that's exactly what I wrote!" and the like. A class period spent going over the exam can be a waste of time. Furthermore, students who have performed very poorly may have trouble paying attention, whereas students who have done very well may be eager to move on.

If you decide to return graded exams in class, you may find it least disruptive to return the exams at the end of the class period. Some instructors post information (such as means or medians) on the board at the beginning of class, to indicate how well the class did as a whole. For example, some instructors write the classes' average on the board during the first class meeting after the exam. This can be a tremendous morale boost if the class has performed well. Naturally, this can raise serious questions in the students' minds if the class has not performed well as a whole. Clearly, there are risks involved in posting this kind of information. In order to prevent students from trying to alter their answers, some instructors announce an "If the exam leaves the room, I can't make any changes to the score," policy before returning papers to students.

An alternative approach is to have students come to office hours, or make short appointments, to pick up their exams. This has advantages over returning the exams *en masse* in class, although meeting with the students individually or in small groups does take quite a long time. The advantages include:

(a) You are able to deal with each student's grade complaints privately, instead of having to explain, or defend, your grading scheme in front of the whole class.

(b) You are able to spend some time with each student working through the parts of the exam that they did not answer well.

(c) You are able to put each students' exam score into context, and advise each student of their options. This can be particularly helpful when you are dealing with students who are bitterly disappointed with their performance on the exam.

(d) If you have an "If the exam leaves the room, I can't make any changes to the score," policy, then there is no opportunity for students to modify their answers on the exam.

4. **Dealing with Students**

This can be very difficult. Some students will be devastated by their exam scores, others may want to have their ego stroked, still others will be adamant that they have been graded unfairly, and demand that you re-grade their paper 'properly.'

Guidelines that are appropriate when dealing with exams are:

• Do not discuss with students how specific individuals performed on the exam. Sometimes, students will ask you how other people in the class did. It is suggested that you politely, but firmly, refuse to answer such questions. Descriptive statistics such as the class mean or median should be enough. Some students may like to know where in the class their exam score placed them (first, second, last, etc.). If you have this information and are comfortable about releasing it, then this is probably all right. It is not all right to tell a student the name of the student who was the first in the class, etc. It violates students' right to privacy to discuss their performance with others.

• If you discuss grades with other instructors, please do not name names.

• Do not discuss your perception of a student's ability in front of other students, whether from your own class or from another class.

• Do not re-grade exams in front of students.

- Do not get into debates over grading policies, or the fairness of exam questions. If students press, then point out that everyone took the same exam, and the same grading scheme was applied to everyone's paper.

If you have students who insist on having their exam re-graded, it is often very helpful to ask the students to write a statement explaining why they deserve more credit. Evaluate their request for more credit based on this statement—usually students who are simply "grubbing" for a higher score will be unable to point to any specific instance where the grading was incorrect, or your grading scheme misapplied. Their statements will reflect this. All of the exams in your class ought to have been graded to the same standard, and re-grading individual exams to a different standard upsets this.

5. **Students Who Cheat After the Exam**

In fairness, it must be pointed out that the overwhelming majority of students are very honest, and would not consider altering their exams. However, almost everyone who has been involved in teaching for any length of time will know of cases where this has almost certainly happened.

One of the problems associated with simply handing exams back to students and asking them to bring grade complaints to you is that it is possible for students to alter their answers, and claim that the exam was graded incorrectly. Unless you have a copy of the exam *before* it was returned to the student, there is no way to definitively prove that the exam was altered.

Copying every single exam is wasteful and unnecessary. A policy along the lines of, "If the exam leaves the room, I can't make any changes to the score," will achieve the same result in almost all cases. If students wish to check through their exam in great detail, then encourage them to meet with you outside of class to do so.

10.7   Suggested Reading

Most of the references that include sections on cheating appear to describe similar student behaviors, and similar strategies for trying to deal with these. All of the references are careful to point out that cheating is a real problem in some courses, and it may not be possible to completely eliminate cheating.

- James Cangelosi. *"Classroom Management Strategies. Third Edition."* [21]

  Chapter 11 ("Dealing With Non-Disruptive Off-Task Behaviors") includes a section on cheating. This book is written to be most immediately relevant to school teachers. Unlike most of the other references, this book speculates on the prevalence and causes of cheating. Mathematics instructors who teach introductory courses may find some of this information helpful. The significant causes of cheating are from students' value systems that may place "test performance" higher than "learning." This may be familiar to instructors who teach courses that students usually take to fulfill some kind of college requirement.

- Bettye Anne Case, ed. *"Responses to the Challenge. Keys to Improved Instruction by Teaching Assistants and Part-Time Instructors. MAA Notes #11."* [23]

  The second part of this book contains a collection of "New Instructors' Manuals" from a variety of institutions. Although somewhat dated, many of these contain advice to new instructors regarding cases of suspected cheating. In particular, these may serve as models for developing a description of your institution's official procedures for handling cheating.

- Wilbert McKeachie. *"Teaching Tips."* [81]

  Chapter 7 briefly discusses some of the ways in which students may attempt to cheat, as well as some ways of preventing and dealing with cheating. In addition to the usual advice, such as having the students "take alternate seats," McKeachie discusses the importance of developing a class atmosphere that has "honesty as a group norm." He also points out that developing instructor-student relationships removes the anonymity that may lead students to believe that they will not get caught cheating. McKeachie suggests the viewpoint that the committees that deal with incidents of cheating provide help to students. McKeachie also suggests that this help may yield long-term rewards for students, rather than simply punishing students for misdeeds.

Chapter 11

Improving Teaching with Graphing Calculators and Computer Algebra Systems

The widespread availability of affordable graphing calculators and computers running powerful computer algebra systems (CAS) is changing the way elementary mathematics is taught. Research in mathematics education (see the suggested readings at the end of this chapter for examples) indicates that the careful use of appropriate technology can help students to achieve learning outcomes that are rare in more traditional mathematics courses. Graphing calculators and CAS have been likened to the microscopes of a biology lab. Just as students can use microscopes to directly observe the working of organisms and other biological systems, calculators and computers can be used by students to explore mathematics. Just as personal observation of the behavior and structure of biological systems can make basic principles of physiology and anatomy more concrete for students, observation of the behavior and structure of mathematical phenomena can make abstract mathematical concepts more concrete for the students.

Nationwide, some instructors have reservations about incorporating such technology into the classroom. Critics of the use of technology in mathematics classes include that the net effect will be that students will fail to learn any mathematics, instead they will simply become adept at using the technology without developing any substantial understanding of any mathematical concepts. Another frequently heard criticism (particularly of calculators) is that the technology will become a "crutch" for students.

These are potentially valid criticisms of the use (perhaps "misuse" is more accurate) of technology in mathematics courses. We wish to point out strongly that we do not advocate the use of technology as a way for students to "short-cut" their learning process by learning some calculator or computer commands by rote. We believe that the use of graphing calculators and CAS in elementary mathematics courses can be a great boost to student learning, but for this to be so the technology must be thoughtfully and carefully incorporated into the course. It is also important for instructors to have some idea of the different kinds of learning outcomes that can be realized through the use of graphing calculators, such as those found in [45]. In this meeting we give examples to demonstrate some of the ways that the use of graphing calculators and CAS can be incorporated into courses. We believe that the examples that we include support and encourage student learning, rather than provide students with ways to avoid learning mathematics.

The most obvious function of a calculator is to do calculations, and of a computer algebra system is to do algebraic manipulations. George Pólya [89] laid out the four steps to solving a mathematical problem as

1. understand the problem,

2. devise a plan,

3. carry out the plan, and

4. look back.

Students of mathematics have traditionally spent a large majority of their time focused on step 3, whereas non-mathematicians who use mathematics in their careers are most likely to spend much of their time on steps 1, 2, and 4, while leaving much of step 3 to a computer. We are not suggesting that manual computation should be abandoned. However, we do believe that the pedagogically sound use of calculators and computers can allow students of mathematics to spend more time involved with steps 1, 2, and 4. In particular, with access to appropriate technology, students can actually spend time solving "real world" problems that involve actual (i.e., possibly messy) data. Calculators and computers can also be used to implement algorithms that, beyond a few iterations, are not very feasible if the calculations must be done by hand.

In addition to allowing students access to accurate and fast calculations, graphing calculators and computer programs can allow students to visualize concepts, explore conjectures, discover theorems, and connect multiple representations of a concept by attacking problems with a variety of methods. Therefore the primary focus of a mathematics course can shift from algebraic manipulations to understanding concepts, developing higher order thinking skills, and working on applications.

There is certainly some amount of cost involved with incorporating calculators and computer systems, as students must learn to use the technology in addition to learning the mathematical content of the course. Therefore, if calculators and computers are incorporated in a manner that does not contribute to student learning, the net effect is adding more (possibly useless) knowledge for the students to acquire.

We have designed this meeting with two purposes in mind. The first purpose is to provide a framework and specific examples within that framework for utilizing the capabilities of graphing calculators and CAS beyond merely as computing devices, algebraic manipulators or table and graph producers. The second purpose is to begin a dialogue among the instructors on some of the subtle issues involved with and the potential drawbacks of incorporating technology, and to begin a discussion on ways to minimize these effects.

11.1 Description and Purpose of the Meeting

This meeting is intended to be a fifty minute session. The primary purpose of this meeting is to help instructors to move beyond simply using graphing calculators and CAS as number calculating, algebra manipulating or curve sketching devices, and start to use them as meaningful teaching tools. After a discussion of the goals of the course, five categories of using technology as a teaching tool will be discussed. Each category will be demonstrated by a specific exercise. Finally, the instructors will have an opportunity to discuss the possible drawbacks and potential dangers of using graphing calculators and CAS.

11.2 Goals for the Meeting

1. To provide instructors with an opportunity to think about how technology can be used to help achieve the goals of the course.

2. To introduce instructors to five categories of activities for using technology as a pedagogical device.

3. To show instructors specific examples of a calculator or computer activity from each of the five categories.

4. To provide instructors with a forum to share ideas that they have produced on using graphing calculators or computers as teaching tools.

5. To provide instructors with an opportunity to discuss how the technology could be used to enhance student learning in upcoming sections of the course.

6. To give instructors the opportunity to think about some of the potential drawbacks of using technology, and how to avoid these situations.

11.3 Preparation for the Meeting

1. Prepare an overhead of the goals of your course.

2. Obtain an overhead projecting calculator or a computer with a projector. (Alternatively, if you plan to hold the meeting in a computer lab, reserve the lab room for the meeting.)

3. Select activities from the end of this chapter to demonstrate during the training meeting (or devise your own).

4. Obtain a program for doing numerical integration (if you plan to demonstrate the Riemann sums activity during the training meeting).

5. Practice the calculator or computer learning activities described at the end of this chapter, and time how long it takes to present each one. Decide how many, and which, of the activities to present.

6. You may choose to invite several experienced instructors to attend the meeting to lead the learning activities.

7. Circulate the agenda for the meeting to all participants well in advance.

8. Make sure the instructors know to bring their calculators to the meeting (if appropriate), or that the meeting will be held in the computer lab (if appropriate).

9. Make copies of meeting materials, and any questionnaires that you plan to distribute, and confirm attendance with participants.

11.4   Agenda for the Meeting (to be distributed to participants)

1. Discussion of the goals of the course, and how graphing calculators and CAS can be used to achieve these goals.

2. An outline of five categories of learning activities that use technology, followed by demonstrated examples of how to run activities that use technology as a teaching tool.

3. Forum for instructors to share ideas on using technology as a teaching tool, most specifically in upcoming sections of the course.

4. Discussion of the potential drawbacks of using graphing calculators or CAS.

11.5   Outline for Running the Meeting

:01–:05 Introduce the meeting, and briefly review the agenda.

:05–:10 Pass out the handout "Technology as a Teaching Tool." Indicate that the goal of this meeting is to move beyond ordinary uses of calculators and computers, and to begin to use their unique capabilities to enhance student learning. The categories of use summarized in the handout are

- **Discovery**—learning by discovery,
- **Algorithms**—using the programming capabilities to expedite calculation algorithms that are tedious by hand,
- **Visualization**—visualizing mathematical notions, such as the tangent line approximation or limits of functions,
- **Solving Problems**—allowing the student to solve problems that without the use of technology would be beyond their current mathematical tools,
- **Connecting Multiple Representations**—connecting graphical, numerical, and algebraic approaches to a situation.

:10–:30 Lead some or all of the following five activities on using technology as a teaching tool. Choose which activities to do based on what the goals of the course emphasize, and how long it takes to present the activities. In order to get through enough activities, present them in a manner appropriate for the level of the audience, but remind the instructors that the in-class use of such activities will take considerably more time. At the end of the activities, there may be a few minutes to brainstorm ideas to add to the lists on the handout, or to discuss resources for finding ready-made in-class calculator activities. A few such resources are the company web sites at www.ti.com., www.hp.com, www.sharp-use.com, and www.casio.com.

The activities outlined in the handout at the end of this chapter are

- Discovering the Fundamental Theorem of Calculus (discovery),
- Approximating Functions Using Polynomials (discovery),
- Evaluating Integrals Numerically (algorithms),
- Visualizing the Derivative (visualization),
- Convergence of Fourier Series (visualization),
- Calculating Extrema in Precalculus (solving problems),
- Using Transforms to Solve Differential Equations (solving problems), and
- Investigating Convergence of Improper Integrals (multiple representations).

:30–:35 Allow the instructors time to share their own ideas on general ways in which technology may be used as a teaching tool, or specific ways that graphing calculators or CAS could be used in upcoming sections of the course.

:35–:45 Lead a discussion on the potential drawbacks of using graphing calculators and computers. Some points that may come out include the following.

- The graphical displays of graphs (especially those displayed on LCD screen with limited resolution) can be misleading if taken at face value. For example, the global behavior of the graph of $y = e^x \cos x$ is difficult to determine on any window. The students may believe that the graph of this function consists of a horizontal half-line on the left side of the x-axis, followed by a series of vertical lines to the right of the origin.

- There can be technical shortcomings. For example, many of the graphing calculators mismanage asymptotes when graphing a function such as $y = \frac{1}{x-1}$, and so the students may believe that the function is continuous at $x = 1$. Another example is that many calculators gloss over discontinuities in graphs of functions such as $y = \frac{x^2-4}{x-2}$. If your course uses examples like this to motivate limits, then this tendency of the calculator may actually distract students from the observation that the function is not defined at $x = 2$. We note that most CAS generally handle these issues better than hand-held graphing calculators.

- Calculators and computers can be a crutch, and cause students to unlearn things that they know (e.g., $\frac{100}{10}$ or $\cos(\frac{\pi}{3}) = \frac{1}{2}$) or to overlook obvious simplifications that can help them to understand the mathematics (for example, students using computer algebra systems to integrate or differentiate functions like $f(x) = \sin^2(x) + \cos^2(x)$).

- Because each student is likely to have his or her own calculator, the machines can be a physical barrier to interaction in the classroom. When working on an activity each student may potentially interact with his or her calculator rather than other students in the class. This problem can be reduced if students normally work with CAS on computers, as two students can normally work at a single terminal.

- Calculator or computer skills are one additional thing to teach on a limited time budget.

- The calculators and especially powerful symbolic manipulation software such as CAS raise tricky assessment issues.
  - Instructors must decide which algebraic skills students should still be able to master.
  - Instructors must decide which types of problems gain importance with technology.
  - The instructors must be aware that the intent of certain problems may be compromised by technology.

– Instructors may feel the need to resort to artificial generality in order to remove the option of using the calculator to do routine computations. However, this works against the intent of integrating technology into the classroom in the first place.

– Calculators raise equity issues when different students own different models of calculators. Problems need to be crafted in such a way as to avoid providing advantages to students with a particular type of calculator. This may also be an issue if students use CAS on their own computers. If students normally work with CAS on computers provided in a lab, then this equity issue may be reduced.

– If students are not allowed to use the technology that they are familiar with when they work on exams, the net effect is more for the students to learn. Alternatively, it can be difficult to schedule a room that is suitably equipped for students to use technology on exams. A third exam-related issue is that many computers are networked which may provide students with an opportunity to either access prohibited references or communicate with others during the exam.

– The storage capabilities of programmable calculators allow them to be used as electronic crib sheets. Likewise, if networked computers are available for exams, then students may have the entire internet at their disposal during the exam.

The instructors may find it interesting to see some of the technological shortcomings of the graphing capabilities. In addition, these can lead to interesting classroom discussions or extra-credit assignments. An easy example is the mismanagement of asymptotes, when graphing a function such as $y = \frac{x^2+1}{x^2-1}$. Another example that works well on graphing calculators is under-sampling of sine graphs. If the sampling rate of the calculator is M, then $N \equiv L \pmod{M}$ implies that the functions $\sin(Nx)$ and $\sin(Lx)$ will agree on all the sample points. For example, observe that the graphs of $y = \sin(6x)$ and $y = \sin(100x)$ for $0 \le x \le 2\pi$ look the same when graphed on the TI-83. For an interesting discussion of under-sampling of sine graphs, see [40].

Follow up the listing of their concerns by addressing some of the issues. Some of the issues may have easy answers, but most will be ongoing concerns that need careful thought on the part of the instructors and the department.

:45–:50 Wrap up the meeting with a word of thanks, and (as appropriate) time for participants' questions, a summary, suggestions for additional readings, or a questionnaire.

11.6 Meeting Materials

1. Copies of the handout entitled " Technology as a Teaching Tool," and an overhead transparency of the handout.

2. Copies of the handout entitled "Learning Activities."

3. A working graphing calculator of the model supported by your department, preferably with an attached overhead projection screen, or a computer with a projector running the CAS normally used in your department. A third possibility is to hold the meeting in a lab room equipped with computers running the CAS.

4. A numerical integration program (if you choose to do the learning activity on Riemann Sums).

Technology as a Teaching Tool:
Sample Activities from Calculus,
Precalculus and Statistics

Discovery

- Effects of changing parameters in function families
- Derivative rules ($\sin x$, x^n, chain rule)
- Fundamental Theorem of Calculus

Algorithms

- Statistical regression
- Newton's method
- Euler's method
- Numerical integration

Visualization

- Global behavior of rational functions
- Relative growth of function families
- Visualizing derivatives
- Visualizing limits of functions

Solving Problems

- Calculating maxima and minima
- Calculating intersections of linear and exponential functions
- Using slope fields to solve differential equations

Connecting Multiple Representations

- Testing convergence of improper integrals
- Linear regression and the correlation coefficient

Learning Activities

1. **Discovering the Fundamental Theorem of Calculus**

   Choose a function with which the students are reasonably familiar (for example a polynomial or a simple trigonometric function) and which has a simple antiderivative. Use the integration capabilities of a graphing calculator or CAS to graph

   $$F(x) = \int_0^x f(t)\,dt.$$

   Then, use the derivative capabilities of the calculator or CAS to graph $F'(x)$ simultaneously with $f(x)$. If the calculator does not have built-in derivative capabilities, you could alternatively define a second function

   $$g(x) = \frac{F(x + 0.001) - F(x)}{0.001}.$$

   Next plot $F(x)$ simultaneously with the known antiderivative of $f(x)$. Repeat with some other functions $f(x)$, and then have the students surmise the statement of the theorem. A sample set of Maple commands for this activity are given below. (The groups of three question marks, "???" indicate places where the user is expected to supply input.)

   ```
   > with(plots):
   > % Enter the function you wish to study as 'f'
   > f:= t -> ???;
   > F:= x -> int(f(t), t= 0 ..  x);
   > plot(F(x), x = ???  ..  ???);
   > plot(f(t), diff(F(t), t), t= ???..  ???);
   > % Enter a formula for an antiderivative of 'f'
   > g:= x -> ???;
   > plot(g(x), F(x), x = ???  ..  ???);
   ```

   Alternatively, if you have access to a slope field program, you can plot

   $$F(x) = \int_a^x f(t)\,dt$$

   for various values of a on the slope field for the differential equation $\frac{dy}{dx} = f(x)$.

2. **Approximating Functions Using Polynomials**

   This activity is designed to show how the idea of a Taylor Polynomial can be "discovered" by successive modifications of the tangent line approximation. Throughout this activity, we will assume $x_0 = 0$ and that $f(x) = \cos(x)$.

   Step 1: Ask the participants to find the equation of the tangent line of $f(x)$ at x_0, and ask the participants to plot both $f(x)$ and the tangent line that they have obtained. Ask the participants to use their plots to find a range of "x-values" where the two plots are indistinguishable.

   Step 2: Point out to the participants that the tangent line approximation is a degree one polynomial. Ask the participants to speculate on whether or not a degree two polynomial would do a better job of approximating $f(x)$. (The question of what "better" means in this context should be something that meeting participants address through their responses.) Ask the participants to experiment with polynomials of the form $q(x) = ax^2 + bx + c$ to find the one that does the "best" job of approximating $f(x)$. Emphasize to the participants a "guessing and checking" (by plotting their polynomial and $f(x)$) strategy, rather than just using the formula for Taylor polynomials. This may be somewhat artificial for the participants, but it will be a realistic approach for calculus students (who don't know Taylor's theorem) to use. Ask the participants to plot both $f(x)$ and $q(x)$ and find the interval of "x-values" where the plots of $f(x)$ and $q(x)$ are indistinguishable.

140 Learning to Teach and Teaching to Learn Mathematics

Step 3: Have the participants calculate,

$$q_1(x) = f(x_0) + f'(x_0)(x - x_0) + \frac{f''(x_0)}{2!}(x - x_0)^2,$$

and notice that this will be similar to the formulas for $q(x)$ that the participants obtained by "guessing and checking" in Step 2. Have the participants speculate on the degree four polynomial that approximates $f(x)$ and have them test their speculations by plotting their degree four polynomial and $f(x)$ and finding the range of "x-values" where the two plots are indistinguishable.

3. **Evaluating Integrals Numerically**

This activity is designed to show how programmable calculators and computers can expedite traditional numerical algorithms that were formerly inaccessible to students beyond a very small number of iterations.

Numerical integration programs for calculators can be found in many calculator instruction manuals, instructor guides for calculus text books and on the web. If no program is available, develop one, or check the calculator's instruction manual to see how to use the summation commands to evaluate Riemann sums. The relevant Maple command is sum. Calculate some integrals numerically with this program.

To expedite the activity, either pass out the program to the instructors in advance of the meeting, or use an overhead projecting calculator. If you are meeting in a computer lab, load your computer program or CAS file into each of the machines before the meeting begins.

This kind of application is usually well understood by instructors. This would be the best activity to skip when pressed for time during the meeting.

4. **Visualizing the Derivative**

This activity is intended to help students to visualize the derivative as the slope of a line that is the limit of secant lines, and connect this geometric interpretation with the symbolic form of the definition of the derivative.

Step 1: Draw a set of axes and a function on the board. Mark the points $(a, f(a))$ and $(a+h, f(a+h))$ on your graph, and join them with a straight line. Write the equation for this line on the board.

Step 2: Write the mathematical definition for the derivative at a,

$$f'(a) = \lim_{h \to 0} \frac{f(a + h) - f(a)}{h}$$

on the board.

Step 3: Ask each person to pick a function to enter into the first function on their calculators or computers. Then have everyone enter

$$y = \frac{f(a + h) - f(a)}{h} * (x - a) + f(a)$$

into the second function on their calculators or computers. The Maple code for this is given below (as before, "???" indicates a place where the user should supply some input).

```
> with(plots):
> % Enter your function as 'f'
> f := x -> ???;
> % Enter numbers for 'a' and 'h'
> a := ???;
> h := ???;
> g := x -> (( f(a+h) - f(a))/h)*(x - a) + f(a);
> plot(f(x), g(x), x = ???  ..  ???);
```

Tell the meeting participants to choose a particular value and store it as a (either in the calculator's memory or define a in the CAS). Tell them to choose smaller and smaller values of h, and graph the functions on viewing windows so that $(a, f(a))$ and $(a + h, f(a + h))$ are visible.

Step 4: Make the point that the derivative formula is the slope part of the equation for the secant line, but with h made extremely small. Remark that the pictures show that when h is made extremely small, the secant line becomes the tangent line. So they should see that a visual interpretation of the derivative is that it is the slope of the tangent line.

5. **Convergence of Fourier Series**

This is a simple activity to address a straight forward issue. Often in courses on differential equations, students encounter Fourier series but develop little or no sense that the Fourier series,

$$f(x) = a_0 + \sum_{n=1}^{\infty} a_n \cos\left(\frac{n\pi x}{L}\right) + \sum_{n=1}^{\infty} b_n \sin\left(\frac{n\pi x}{L}\right), \tag{11.1}$$

is anything beyond a collection of symbols that appear in the solution of a differential equation. This activity uses a CAS to plot graphs of the partial sums of a Fourier series, along with a graph of the function that these partial sums approximate. In this activity, the function that will be approximated is $k(x) = 1 - \mid x \mid$ over the interval $[-L, L]$ $(L > 0)$.

Step 1: Present the participants with the formulas for determining Fourier coefficients,

$$a_0 = \frac{1}{2L} \int_{-L}^{L} k(x) dx, \tag{11.2}$$

$$a_n = \frac{1}{L} \int_{-L}^{L} k(x) \cos\left(\frac{n\pi x}{L}\right) dx, \tag{11.3}$$

and,

$$b_n = \frac{1}{L} \int_{-L}^{L} k(x) \sin\left(\frac{n\pi x}{L}\right) dx. \tag{11.4}$$

Ask the participants to set up functions using their CAS that they can use to evaluate the Fourier coefficients of $k(x)$. Maple commands that would be suitable for this are given below.

```
> k := x -> 1 - abs(x);
> a := n -> (1/L)*int(k(x)*cos(n*Pi*x/L), x=-L..L);
> b := n -> (1/L)*int(k(x)*sin(n*Pi*x/L), x=-L..L);
```

Step 2: Ask the participants to choose a value for the positive constant L, and to set up a function that will give the "Nth" partial sum of the Fourier series, (11.1). A suitable Maple command would be:

```
> L := ???;
> m := (N, x) -> a(0)/2 + sum(a(n)*cos(n*Pi*x/L), n = 1..N)
+ sum(b(n)*sin(n*Pi*x/L), n = 1..N);
```

Step 3: Ask the participants to plot $k(x)$ and $m(N, x)$ for various values of N and different ranges of "x" values. While the participants do this, have them use their experiments to answer the questions:

- How does the value of N affect your plot?
- Over what range of "x" does $m(N, x)$ do a good job of approximating $k(x)$? What does $m(N, x)$ do outside of this range of values?
- Are there any points where $m(N, x)$ seems to have a lot of trouble approximating $k(x)$?

6. **Solving Extremal Problems in Precalculus**

This activity is designed to introduce precalculus students to complex modeling situations, and allow them to solve extremal problems before they have access to the derivative.

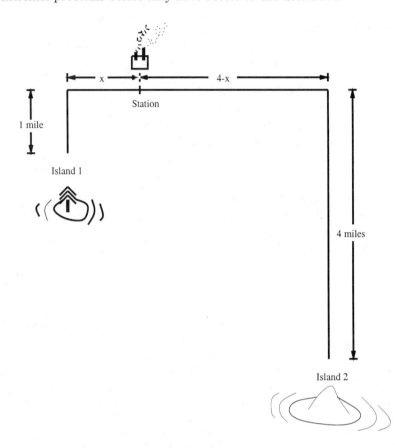

Problem: There are two islands off the coast of Maine. One island is one mile straight off the shore, while the other island is four miles straight off the shore. If measured straight along the shoreline, the islands are four miles apart. The electric company wants to build a station on shore to supply both islands with electricity.

Where along the shoreline should the station be built in order to minimize the sum of the distances from the station to each island?

Solution: Once the students attain the model

$$D(x) = \sqrt{x^2 + 1} + \sqrt{(4 - x)^2 + 16},$$

they can use the graphing calculator or computer to find the distance, x, which minimizes the sum of distances, $D(x)$.

7. **Using Transforms to Solve Differential Equations**

The Fourier transform is a familiar device from courses in analysis and signal processing. One application of the Fourier transform (solving differential equations) is frequently underplayed in differential equations courses. This is (at least in part) due to the fact that Inverse Fourier transforms are notoriously difficult to calculate by hand. In this activity, participants are asked to solve a relatively innocent-looking differential equation by hand, and then to use a CAS to both find the solution using Fourier transforms and to verify that the formula found actually does satisfy the differential equation.

The differential equation for this activity is:

$$\frac{du}{dt} - u = e^{-t^2}. \tag{11.5}$$

Step 1: Present the differential equation to the participants, and ask them to find a solution of the equation by hand. (For example, by solving the homogeneous equation and then trying to find a particular solution.)

Step 2: After participants have struggled for a while, ask them if they know how the Fourier transform of $\frac{du}{dt}$ is related to the Fourier transform of $u(t)$. Ask the participants to use this information to find an expression for the Fourier transform of a solution of the differential equation, (11.5). Some useful Maple commands (or their equivalents in the CAS that your institution uses) are:

```
> with(inttrans):
> fourier(exp(-1*t^2), t, omega);
```

Step 3: Ask the participants to explain how they could use the expression that they found in Step 2 (at least in theory) to find a solution of the differential equation, (11.5). Ask the participants what practical difficulties they see to accomplishing their plan.

Step 4: Show the participants how to compute an inverse Fourier transform using your CAS. The Maple command for computing the inverse Fourier transform of $\frac{1}{1+\omega^2}$ is:

```
> invfourier(1/(1 + omega^2), omega, t);
```

Step 5: Ask the participants to use the differentiation commands of the CAS to verify that the formula that they have obtained from Step 4 to check that this formula actually does satisfy the differential equation, (11.5).

Step 6: Ask the participants to use the results from the CAS in Step 4 and the solution of the homogeneous equation from Step 1 to construct a general solution for Equation (11.5).

8. **Investigating Convergence of Improper Integrals** [1].

This activity is designed to exploit the capabilities of the graphing calculator to relate the algebraic, numeric, and graphical approaches to investigating the convergence of improper integrals.

Step 1: Compute the integrals $\int_1^{10} f(x)\,dx$, $\int_1^{1000} f(x)\,dx$, and $\int_1^{100000} f(x)\,dx$ for $f(x) = \frac{1}{x^2}$ and $f(x) = \frac{1}{\sqrt{x}}$ using Riemann sums, computer integral commands or the Fundamental Theorem of Calculus. Observe that the integrals of $\frac{1}{x^2}$ are approaching 1, while the integrals of $\frac{1}{\sqrt{x}}$ are growing larger and larger.

Step 2: Graph the functions $y = \frac{1}{x^2}$ and $y = \frac{1}{\sqrt{x}}$ using several different ranges of plotting values or sizes of graphing windows. Remind the students of the global behavior of the functions first, and then use computers or calculators to graph them on the window $0 \le x \le 10$ by $0 \le y \le 1$, on the window $0 \le x \le 1000$ by $0 \le y \le .1$, and then on the window $0 \le x \le 100000$ by $0 \le y \le 0.01$. Have the students notice how dramatically different the behaviors of the two functions are becoming. In fact, on the last window the graph of $y = \frac{1}{x^2}$ does not even appear. Have the students play around with some other scales, and conclude that the tail of $y = \frac{1}{\sqrt{x}}$ is substantially fatter than that of $y = \frac{1}{x^2}$.

Step 3: Connect the graphical and numeric viewpoints. For example, graph $y = \frac{1}{\sqrt{x}}$ on the window $0 \le x \le 1000$ by $0 \le y \le 0.1$. Have the students estimate that the area under the curve covers at least $\frac{1}{3}$ of the viewing window, and so accounts for at least 33 units of area. Alternatively, use the trace feature to find that the y value always stays above .3 on this window, so that a certain lower bound for the area under the curve in this window is 30. Repeat on the window $0 \le x \le 100,000$ by $0 \le y \le 0.01$. As before, the area under the curve seems to cover at least $\frac{1}{3}$ of the viewing window, and so accounts for at least 333 units of area, and certainly covers at least 300 units of area. If the same process is attempted with $y = \frac{1}{x^2}$, the function dies out and defies any reasonably large lower bound on such viewing windows. The conclusion is that the fact that the tail of a function gets infinitely thin is not enough to guarantee that the area beneath the curve is finite. If there are long stretches where the tail does not decrease very rapidly, the area under the curve can be quite substantial. However, if

---

[1]Adapted from [5]

a tail dies out rapidly enough, this will not be the case, because the tail thins out to nothing before the area under the curve can become substantial.

Step 4: Use algebraic methods to calculate which of the integrals $\int_1^\infty f(x)\,dx$ converges, and the value of the convergent integral.

Step 5: Explain how observations made about the numerical and graphical behaviors of the functions provide a way of understanding the algebraic results.

Step 6: Use these functions to study the reverse situation: whether or not all vertical asymptotes are integrable. For example, which of the following integrals represent a finite area?

$$\int_0^1 \frac{1}{\sqrt{x}}\,dx \text{ or } \int_0^1 \frac{1}{x^2}\,dx$$

11.7 Suggested Reading

There is a lot of material available concerning the use of technology in teaching mathematics. We have grouped the suggested readings into three categories: general references, instructional materials and research.

General References

- L. Carl Leinbach, et. al., eds. *"The Laboratory Approach to Teaching Calculus. MAA Notes# 20."* [74]

  The introduction to this collection of articles begins by noting that the term "Laboratory Calculus" "... conjures up the image of a room filled with microcomputers loaded with the instructor's choice of software." Although this reference is almost ten years old, it contains descriptions of well-known implementations of technology in teaching calculus. The book is divided into three parts: general issues, examples of established programs and calculus laboratory projects.

- Zaven Karian, ed. *"Symbolic Computation in Undergraduate Mathematics Education. MAA Notes# 24."* [67]

  Like the previous reference, this is an edited collection of articles. The articles in this book are specifically concerned with the use of computer algebra systems (CAS) in undergraduate mathematics courses. The collection is divided into five sections: Pedagogical issues, Symbolic computation in calculus, Symbolic computation in linear algebra and differential equations, Symbolic computation in advanced courses, and Getting started. The first few articles raise (and speak to) serious issues in the use of CAS and the assessment of courses that employ technology. Many of the articles include numerous examples of projects and activities using CAS. The last two articles are a survey of some of the literature (circa 1992) and an annotated bibliography of using technology in teaching undergraduate mathematics.

- M. Kathleen Heid. "Calculators on Tests—One Giant Step for Mathematics Education." [58]

  This article argues that calculators should be used for testing, as well as for instruction. The article argues that as a result of using calculators on tests, mathematics curricula would shift from computation to deeper development of concepts and to mathematical modeling. The suggestion is that a broader curriculum would result.

- Boyce, William E. and Joe Ecker "The Computer-Oriented Calculus Course at Rensselaer Polytechnic Institute." [18]

  This article documents the experiences and conclusions of a long-range program to revitalize undergraduate instruction in mathematics utilizing the computer algebra system, Maple. The article includes two examples of computer-intensive calculus laboratory projects.

- Judson, Phoebe T. (1991) "A Computer Algebra Laboratory for Calculus I." [66]

  This article describes a weekly computer algebra laboratory course for Calculus I students during the 10-week spring semester, 1990, at Trinity University. The article offers advice on the establishment and management of this type of laboratory including descriptions of student participation trends and student reactions to the laboratory course.

Instructional Materials

- Glendon Blume and M. Kathleen Heid, eds. (1997) *"Teaching and Learning Mathematics with Technology."* [13]

  The chapters of this compilation are replete with classroom-tested ideas for using technology to teach new mathematical ideas and to teach familiar mathematical ideas better. Of particular relevance to the forms of technology that we have described are: Chapter 1, "Using the Graphing Calculator in the Classroom: Helping Students Solve the "Unsolvable"; Chapter 5, "Composing Functions Graphically on the TI-92" and Chapter 9, "Mathematically Modeling a Traffic Intersection."

- Frank Ward and Doug Wilberscheid. *"Insight Into Calculus Using Texas Instruments Graphics Calculators."* [117]

  This book represents a collection of projects for students to complete using graphing calculators. The authors state that they do not wish to remove the need for a solid conceptual understanding of calculus. Instead, they claim that graphing calculators can provide insights into the concepts of calculus. Some of the projects assume that students have access to a calculator based laboratory (CBL) system in addition to graphing calculators.

- The "Technology Tips" column (and others) of the NCTM publication "Mathematics Teacher" regularly includes detailed descriptions of mathematical activities that employ graphing calculator or computer applications.

- The manufacturers of the graphing calculators popularly used in college courses often have activities and programs available on their web sites. In addition, several textbooks have supplements that provide calculator based projects.

Research

- Penelope Dunham and Thomas Dick. "Research on Graphing Calculators." [45]

  The studies cited in this article, for the most part, refer to the use of graphing calculators by students in high school and college courses. The mixed results of comparisons of student achievement between groups with and without graphing calculators are presented. Although there is no clear picture for student achievement overall, results are presented where there is convincing evidence. These results show that students who use graphing calculators (1) were better able to relate graphs and equations, (2) were better able to read and interpret graphical information, and (3) have an increased "example base," etc.

- M. Kathleen Heid "Resequencing Skills and Concepts in Applied Calculus Using the Computer as a Tool." [59]

  This article describes a research project conducted during the first 12 weeks of an applied calculus course. In this study, two classes of college students studied calculus concepts using graphical and symbol-manipulation computer programs to perform routine manipulations. Three weeks were spent on skill development. The article presents evidence to suggest that students showed better understanding of concepts and performed almost as well on routine skills, compared to students who had spent the entire semester studying applied calculus without using computer programs for routine manipulations.

- Phoebe Judson "Elementary Business Calculus with Computer Algebra." [65]

  A common perception (particularly among instructors who are committed to a traditional view of teaching) is that the acquisition of basic mathematical skills (such as performing algebraic manipulations) must occur before the development of higher-order mathematical skills. This article describes various ways that the computer algebra system Maple was used to facilitate the resequencing of the acquisition of basic skills and applications of mathematics within an elementary college-level business calculus course. Experimental results suggested that skills acquisition is not a prerequisite to conceptual understanding or problem-solving ability.

- Robert Mayes "The application of a computer algebra system as a tool in college algebra." [79]

  This article presents a comparative analysis between traditional and experimental college algebra courses. The study concentrates on the effects of using computers on students' problem solving ability and the effects of the computer-intensive environment on computation and manipulation skills.

- Jeanette Palmiter "Effects of Computer Algebra Systems on Concept and Skill Acquisition in Calculus." [87]

  This study compared the performance of two groups of university students. One group was taught calculus using a computer algebra system, whereas the other used paper-and-pencil computations. The computer group scored significantly higher than their counterparts on both a conceptual knowledge test and a test of calculus computational skills.

- Laurel Cooley "Evaluating Student Understanding in a Calculus Course Enhanced by a Computer Algebra System." [32]

  This article describes an experimental study in which two sections of calculus were taught using the same materials, except one section was enhanced with the computer algebra system Mathematica. The article presents results that suggest that the students in the technology group had advantages in understanding certain key topics in calculus such as limits, derivatives, and curve sketching compared to the other section.

Chapter 12

Making Lesson Plans

An obvious skill that needs to be acquired by new instructors is the ability to plan an effective lesson. In a course consisting primarily of lectures, this translates into writing clear and engaging lectures that include ways of assessing the lectures' effectiveness. There has been much written on the art of preparing such lectures. On the other hand, in an interactive student-centered [1]course, planning an effective lesson translates into a very different activity. The instructor must decide how to balance the various teaching methodologies at his or her disposal so as to achieve the different learning goals set forth for that session. Allowing for a combination of approaches provides the opportunity for a richer learning environment, but it also demands additional considerations such as which methodologies to employ in different parts of a class, how to manage the different activities planned, and how to develop connections among the activities into a coherent lesson.

One criticism that advocates of a purely lecture course make of student-centered courses is that students might actually spend less time out of class engaged in the material as a result of having engaged in active learning in the classroom. In fact, we do not assume that our students are working less outside of class merely because they are working differently inside of class. Instead, the model advocated here requires the students to read the material before coming to class for an initial level of understanding (see Chapter 5). In addition, classroom lessons are designed so that students leave class with a more mature but still developing understanding. Students are then expected to deepen and strengthen their understanding by further reflection on the reading, review of their class notes, and practice doing individual and team homework problems.

There are many possible reasons for using student-centered instruction. Most, if not all, mathematics instructors (even those with a traditional conception of the teaching and learning environment [37]) believe that students learn mathematics by *doing* mathematics. Therefore, it is highly sensible to structure a mathematics lesson so that students actually do mathematics in class. In addition, activities that provide opportunities for verbalizing, discussing, and writing about mathematics can facilitate and solidify student understanding. An interactive classroom can provide more frequent and direct opportunities for assessment and feedback, and through this assessment and feedback the instructor can find opportunities to directly manage student engagement with the material. Finally, the varied teaching methodologies that are utilized in an interactive classroom can be more engaging for students, especially those of the internet generation.

New instructors' natural conceptions of learning and teaching mathematics often make them instinctively emphasize instructor-centered activity [37]. In this meeting we provide a guided lesson planning worksheet that has been used to make it possible for new instructors to include more student-centered activity in their lessons [39]. We should emphasize that the worksheet is intended as an introduction to many of the considerations involved in planning an interactive lesson, and as such it requires the instructors to conceptualize many issues that should eventually become second nature to them. Therefore, if the worksheet fulfills its function, it should make itself obsolete as the instructors move to internalizing many of the processes involved.

---

[1]We do not mean to imply that instructors who employ only the lecture method are de facto not centering their lectures upon what they hope the students will learn. By a student-centered course, we mean a course that combines many methodologies, most of which actively engage the students within the classroom, such as collaborative learning, student presentations, interactive lectures, direct questioning, and writing.

Whether one incorporates a student-centered style of instruction or a lecture-based method, it is paramount that the class lesson be prepared with student learning goals in mind. It is much too easy to choose several problems from the book to have the students work on in class, or to give a lecture based upon a section of the textbook, without giving thought to what the students should be able to do or to know once the lesson is completed. Thus, in this meeting we begin by asking the instructors to identify the main points of the day's section and to translate these into learning goals that are deliberately student-centered. We then ask them to make a lesson specifically constructed in order to achieve these objectives. Because different kinds of mental processes appear to be best facilitated by different kinds of student activities and different kinds of instructor activities, we encourage the instructors to use varied teaching methodologies to achieve the identified learning goals.

Just as any honest lecturer will allow that the lecture method can be entirely ineffective when done poorly, we know that there are many valid criticisms of a student-centered method of instruction when it is badly implemented. One large source of trouble is mismanagement of cooperative learning activities, and we have addressed this in Chapter 4 of this book as well as elsewhere [38]. Another potential problem is that allowing students to have partial control over the flow of the classroom activities provides extra opportunities for disruptive student behavior, and this problem is addressed in Chapter 9. In addition, there are more subtle issues inherent in a student-centered classroom that we hope to address through this meeting. Because most of the activity in the classroom involves student participation, it can be more difficult than in a lecture setting for instructors to find opportunities to share their expertise or insight on the material. Because the instructor shares some of the control over classroom activities with the students, it can be more difficult to manage time than in an uninterrupted flow of lecturing. Finally, because there are disparate activities interspersed throughout the class period, more care needs to be taken in synthesizing and summarizing the content into a coherent whole. Therefore, this meeting asks instructors to carefully consider the different parts of a lesson, how to relate them to each other, how to plan for time management, and how to wrap up individual activities and the lesson as a whole.

12.1   Description and Purpose of the Meeting

This meeting does not have a definite time frame, and the length of the meeting will depend upon the working speed of the participants. It revolves entirely around a worksheet designed to walk new instructors through the stages of planning an interactive lesson. The purpose of the meeting is to introduce the instructors to one method of planning lessons, while they actually generate a lesson plan for their next class session.

12.2   Goals for the Meeting

1. To introduce the instructors to one method of generating lesson plans for an interactive classroom.

2. To provide facilitation and assistance as the instructors generate a lesson plan for their next class period.

12.3   Preparation for the Meeting

1. Plan a lesson using the lesson planning worksheet in order to familiarize yourself with the process.

2. Make enough copies of the lesson planning worksheet.

3. Ask the instructors to read ahead in their textbooks through the section that they will be covering in the next session.

4. Ask the instructors to bring their textbooks to the instructor meeting.

5. Notify the instructors that they will be planning their next lesson in this meeting, so that they will think in terms of a real, rather than an idealized, class session.

6. Circulate the agenda for the meeting to all participants well in advance.

7. Make copies of meeting materials, and any questionnaires that you plan to distribute. Confirm attendance with participants.

12.4 Agenda for the Meeting (to be distributed to participants)

1. Generation of a lesson plan for the next class session using a lesson planning worksheet.

12.5 Outline for Running the Meeting

:01–:05 Introduce the meeting, and explain what the instructors will be doing with the lesson planning worksheet.

:05–:45 Divide the instructors into groups of three. Pass out to each instructor a copy of the lesson planning worksheet. Ask the instructors to work through the worksheet in their small groups. Perhaps the most delicate, and sometimes ignored, aspects of planning an interactive lesson are identifying the goals for the day, deciding on a strategy to address each goal, and managing the overall flow of the class. Therefore, when you see that most groups have finished Step 2, call attention to the front of the room, and facilitate a short discussion of the goals for that section based on the responses of a few of the instructors. Similarly facilitate a large-group discussion after you see that most groups have finished Steps 3 and 4. Continue to float around the room assisting instructors, and answering questions as necessary. If instructors finish early, tell them that they are free to leave once they've completed the worksheet and the questionnaire, if you decide to use one.

:45–:50 Wrap up the meeting with a word of thanks, and (as appropriate) time for participants' questions, a summary, suggestions for additional readings, or a questionnaire.

12.6 Meeting Materials

1. Copies of the handout entitled "Lesson Planning Worksheet."

2. Copies of the course textbook.

Lesson Planning Worksheet

Step 1: Read the section to be covered.

Typically, you will not need to read the section in as painstaking depth as you do your own textbooks, or as you hope your students do with this book. However, it is very important that you know what topics are treated in the section, and how the book treats them. You will plan your lesson under the assumption that your students have read the section. This is impossible to do unless you have read it yourself.

Step 2: Identify the learning objectives for the lesson.

After you have read the section, jot down the 3–5 main ideas that the book presented. Write the main ideas for this section here. Add any other main points that you want to address, that the book did not.

1.

2.

3.

4.

5.

Once you have identified the main ideas from the section, use them to formulate the 3–5 student learning objectives for the class session. These objectives should state what bits of knowledge and what skills students should have acquired by the end of the lesson. Make sure that the objectives focus on the students' learning, and are not instructor-focused. This will help you to concentrate on their learning, and not on your performance. Be careful to state these objectives in a way that will allow you to observe whether or not the students have actually achieved each objective. Write the learning objectives for this section here.

1.

2.

3.

4.

5.

Step 3: Decide a strategy to address each goal.

In an interactive classroom you will be seeking a balance between lecturing and student-centered activities. A model lesson will likely consist of an interplay between short mini-lectures and activities that either foreshadow or solidify the ideas addressed in these mini-lectures. After identifying the goals, you will want to decide which goals will be best addressed through an instructor explanation, which will be best addressed through a student activity, and which may require both. Keep in mind that a mini-lecture can include some student participation, and an activity will almost surely include some instructor explication, so these strategies aren't completely dichotomous. Look back at your goals from Step 2, and identify what type of strategy, or strategies, you will use to address each by marking "lecture," "activity," or "both."

1.

2.

3.

4.

5.

Step 4: Begin time management.

Now that you know very roughly what you will be doing, you need to decide a time frame in which to work. Identify what administrative issues you have on your agenda, and how long to allocate for each. Then identify the intended length of each of your mini-lectures, by dividing them into 5-minute lectures and 10-minute lectures. To maximize student attentiveness, do not lecture for more than 10 minutes at a time without switching gears. Finally, subtract from the total class time to find the amount of time remaining for student activities. Use this table to plan for your next class.

Activity	Time allotted
Going over HW/Quiz	
Quiz	
Administration	
Mini-lecture 1	
Mini-lecture 2	
Total allocated	
Minutes remaining	

Step 5: Planning the student activities.

Now you will decide what activities the students will do in the lesson. Recall from Step 3 how many activities you had in mind, and from Step 4 how much time you have at your disposal. Choose an activity for each goal from Step 3 that required one. To begin with, it may be easiest to choose a problem out of the textbook that addresses the stated goal. At other times you may design your own activity by making up a problem, designing a worksheet, or creating some other activity that reinforces a skill (e.g., roundtable activities or checking each others individual work in pairs). Designing your own activities can be more interesting, rewarding, and difficult. Because of time considerations, consider for this section only problems from the textbook. Try to vary using activities done in groups, in pairs, and individually. Write down the activities you choose for this section, and whether you will do them in groups, pairs, or individually.

Goal (step 3)	Problem	Grouping

Now decide how long to allocate to each activity. Group activities often take longer than you hope, but having a particular time-frame that you clearly communicate to the students can help you manage the time efficiently. Make sure you allocate approximately two minutes each to the set-up and wrap-up of the activity. When looking at problems that are especially difficult, or that have many parts, it is often easier to plan and manage the time if you break the time-frame down into pieces. For example you could allocate 4–5 minutes for parts (a) and (b) of a problem, then take two minutes to discuss these parts with the class. After this you would have the class continue with parts (c) and (d), etc. Write down how long you expect to spend on each of the activities for this section, and make notes on how you can break down the longer problems.

Problem	Time	Breakdown

Finally, make notes on the particular execution of each student activity. Some considerations are what you will say to set up the problem, the mechanics of getting them into their groups and working, how you will ask for the results to be reported, what you will want to emphasize, and how the main points of the activity relate to the previous or upcoming lectures. Make notes on the activities for this section here.

Step 6: Writing scripts for the mini-lectures.

Go back to Step 3 to recall the goals you plan to address via mini-lectures. Write out a "script" for each mini-lecture on note cards or paper that you can carry around with you. When writing your scripts

- anticipate students' difficulties with the material and how you will help them (plan how you will simplify the language and give intuition),

- move from concrete to abstract,

- plan what you will write on the board, especially in the way of examples and graphs,

- decide on a higher-level question that you can ask your students during the mini-lecture, and

- plan what you will say during the transition from the lecture to an activity or from the lecture to another lecture.

Write a script for each mini-lecture of this section on the next page.

Mini-lectures

Step 7: Looking back.

Refer back to the 3–5 learning objectives that you identified in Step 2. Now you will plan how to briefly assess the effectiveness of your chosen strategies in meeting these learning objectives. If you have been properly facilitating the group activities, then you should already have an idea of how well the students are meeting the learning objectives. You need to also identify how you will assess their understanding of the mini-lectures and individual activities that you have given them. Some brief assessment techniques that can be used during class are to call on a few students with questions, to give the class a question and ask them to vote on the answer, or to have them try a practice problem that they trade with their neighbor, and then find out what portion of the students successfully completed the problem. You may also consider an overall assessment of the class period. As an example of this kind of assessment, you can end class by collecting a "minute paper," in which you ask students to summarize what they learned, to tell you their muddiest point, or to do a brief exercise.

Write down how you will assess whether or not the learning goals that you identified in Step 2 have been met by most of your students.

1.

2.

3.

4.

5.

If you find that the students have not succeeded in obtaining the skills or knowledge that you intended for them to, then you will have to decide how to alter the course of your plans in order to revisit the issue, perhaps in a new way. Adjusting your plans "on the fly" is, of course, a very difficult proposition. However, it is a skill that you should work to acquire as you become more experienced, and it is a possibility of which you should be aware.

Step 8: Putting it all in order.

Write down an outline of the order of the day's administration, lectures, and activities.

Finally, jot down how you will wrap up the class period. Try to regularly include a short summary of what the students' learned that day or a preview of the upcoming reading. Jot down your notes on how to end the next class period.

Final note: It is better to over-plan than to under-plan.

Often you will plan too much, and you will be unable to get through your entire day's lesson. In these instances, assign the activities that you were unable to attend to as extra individual homework, or resort to emergency lecturing if you get extremely behind. As you gain experience, you will hone your sense of how much to include in a class period.

12.7   Suggested Reading

- Matthew DeLong and Dale Winter. "An Objective Approach to Student-Centered Instruction." [39]

  This article presents the algorithmic approach to lesson planning captured in the worksheet of this chapter along with a schematic for understanding this approach to lesson planning. Research literature on lesson planning in mathematics is reviewed and common lesson planning practices described. The concept of *student learning objectives* is explicitly defined. The article includes two sample lessons (one from calculus, one from linear algebra) planned using this procedure. Finally, the article describes some ways in which the lesson planning practices of novice instructors using the lesson planning worksheet differ in highly significant ways from the kinds of limiting practices that were their natural tendencies.

- Matthew DeLong and Dale Winter. "Novice Instructors and Lesson Planning." [37]

  In this article the authors study a cohort of novice college mathematics instructors to determine what they did to plan lessons, what patterns of content or activities were included in their lessons, how they planned to assess the effectiveness of their lessons, and what factors influenced their lesson planning practices. It was apparent that the novice instructors' own experiences as learners were very influential in their initial approaches to planning lessons, and that the novice instructors had little to no awareness of the cognitive activity of students. These factors led the instructors to believe that their primary role was to organize content and transmit knowledge (hence lecture predominated and student activities were sparse), to believe that their primary objective was to cover the material, to leave considerations of assessment out of their thinking when planning and implementing lessons, and to believe that the most effective student learning occurred outside of class

- Margaret A. Farrell and Walter A. Farmer. *"Secondary Mathematics Instruction: An Integrated Approach."* [46]

  This is a textbook for pre-service secondary mathematics teachers. The approach to lesson planning developed in Chapters 1–8 is similar to the one we advocate. The authors define and categorize instructional objectives in Chapter 5.

- Bonnie Gold, Sandra Z. Keith, and William A. Marion, eds. *"Assessment Practices in Undergraduate Mathematics. MAA Notes #49."* [55]

  This book reports on a large variety of assessment practices in mathematics used by more than 100 contributors. Although the scope of assessment is broad, including assessing individual teachers and departmental majors, there is a wealth of ideas in Part II on assessment in the individual classroom. This could be used as a resource for different types of classroom activities and assessment methods when working through Steps 5 and 7 of the lesson planning worksheet.

Chapter 13

Strategies for Motivating Students

A training program for new instructors will naturally begin with attention to day-to-day concerns. It is only after some familiarity and facility is achieved in these areas that a novice instructor has the opportunity for reflection on what is working in the classroom and why. One area for consideration at that point is the subject of student motivation.

An instructor's approach to student motivation will likely depend upon where he or she is situated on the pragmatist-idealist continuum. A total idealist would believe that the inherent beauty and power of mathematics should lead to intrinsic student motivation to learn the subject matter. A total pragmatist would maintain that the extrinsic motivation of grades alone determines a student's approach to mathematics. It would clearly be a superior situation if all students were intrinsically motivated to learn. This, unfortunately, is not the current reality of college mathematics courses, nor is it likely ever to be. In the absence of the ideal, an instructor must give careful consideration to the interplay between intrinsic and extrinsic motivators and the role that the instructor plays within each. Care must be taken, because under certain conditions extrinsic rewards can damage intrinsic motivation.

Student motivation is a complex and varied issue for which no one has all the answers. However, much can be learned by defining terms and comparing and contrasting possible approaches. Therefore, the bulk of this meeting is devoted to introducing the instructors to definitions of intrinsic and extrinsic motivation, and to analyzing some of the strategies that fall under each heading. In particular, the following strategies for motivating students are discussed: grades, subject matter, excellence, success, and goal setting. After a review of these strategies, instructors are asked to reflect upon their own views of what it means to motivate students and how they may incorporate some of these motivational strategies in order to meet goals that they have set for themselves and their students.

The remaining portion of this meeting is devoted to a discussion of a hindrance to motivation that seems especially peculiar to mathematics: performance anxiety. Tobias [113] has found that most average college students do not lack the cognitive abilities necessary for success in elementary college mathematics courses, but that many of these students believe that they lack these abilities. Like motivation, math anxiety is a complicated subject that cannot be adequately addressed in a short section of one meeting. However, the discussion introduces some simple suggestions that serve as a starting point for addressing this issue in the instructors' classes.

13.1 Description and Purpose of the Meeting

This meeting is intended to be a fifty minute session. The meeting is intended to provide some strategies for instructors to use to motivate the students in their class, an appreciation for different aspects of motivation, and a forum for experienced instructors to share some of the techniques and strategies that they have had success with in the past.

The hope is that meeting participants will gain insights into the different motivational processes or strategies that they can employ with their students, the kinds of situations in which each strategy may be helpful, and the kinds of long-term and short-term results that they may reasonably expect to see. It is also hoped that meeting participants will gain a wider appreciation of the nature of the motivational process, and of their role in the different strategies.

The meeting also includes a discussion on math anxiety, how it affects student motivation, and some ideas for addressing this issue in the instructors' classes.

13.2   Goals for the Meeting

1. To provide instructors with ideas that they can use to motivate students in their own teaching situations.

2. To provide instructors with an opportunity to think about what they would like to motivate their students to do and how they can go about doing that.

3. To provide instructors with a general description of several motivational strategies, concentrating on their strategies as processes in which the instructor is a participant, rather than as actions that the instructor has directed the students to perform.

4. To provide instructors with an opportunity to listen to and ask questions about examples of motivational strategies that other instructors have tried in their classes, and the results that they have observed.

5. To provide instructors an opportunity to think about ways that they can increase student confidence and decrease student math anxiety in their classes.

13.3   Preparation for the Meeting

1. Check with experienced instructors to see if any have addressed the issue of student motivation. Try to find out what it means to them to motivate their students, and what kinds of actions they take to achieve this. Ask them about the results they have noticed, and whether or not, in their opinion, the actions they took actually had the desired results. If not, you could ask them what they plan to do differently in the future.

2. Invite two or three experienced instructors who have something to say on student motivation to participate in the meeting by preparing three-minute talks on the following.

   - What does it mean to motivate students?
   - What actions have I taken to try to motivate the students in my classes?
   - What results did I see?
   - What would I do differently in the future?

3. (Optional—although this can really make the meeting much more useful for all participants.) Ask as many meeting participants as you can what they think it means to motivate their students, and what, if any, concerns they have about the levels of motivation in their classes. Ask them if they have taken any steps to address possible problems, and if so, how this has worked out.

4. (Optional—although this can really make the meeting much more useful for all participants.) Share with your experienced speakers the needs that meeting participants have expressed to you and ask that presentations address those needs as closely as possible. Tailor the handouts to support such presentations.

5. Circulate the agenda for the meeting to all participants well in advance.

6. Make copies of meeting materials, and any questionnaires that you plan to distribute, and confirm attendance with participants.

13.4 Agenda for the Meeting (to be distributed to participants)

1. Outline of some general ideas of what it means to motivate students, and some ideas for implementing these ideas in your classrooms.

2. Description of some strategies that people in our department have tried over the years, and the results that they have seen from their actions.

3. Opportunity to examine exactly what it means to you to motivate students, and generate ideas for what kinds of actions you can take.

4. Review and discussion of ideas of what it means to motivate students, and some of the ideas for how to do this.

5. Discussion of math anxiety and some ideas for addressing it.

13.5 Outline for Running the Meeting

:01–:05 Introduce the meeting, and briefly review the agenda.

:05–:20 Ask experienced instructors to give their talks on their perceptions of what it means to motivate students, their efforts to do this, and the results that they have seen. The meeting leader's role here is to keep things moving, to try to avoid the meeting bogging down in arguments over minor details, and to make sure that ideas are considered positively as well as examined critically. There may be more time here than is strictly needed, unless you really have a lot of experienced instructors with valuable advice, so this extra time can be used for questions and answers.

:20–:30 Distribute the handout on different motivating strategies, and give participants a chance to look them over. Provide an overview or summary of what you see as the most important techniques. A way of organizing this overview may be to highlight one relatively short-term, short-duration, low-commitment strategy, and contrast this with what is involved in a longer-term, longer-duration, higher-commitment strategy. Meeting participants should be able to develop a clear idea of their place as instructors in these strategies, and they should get a sense of what enacting each strategy would demand of them personally.

:30–:40 Lead a discussion on math anxiety, how it affects student motivation, and some ideas for addressing this complex issue. Math anxiety is caused by a student's belief that he or she lacks the cognitive abilities necessary for success in mathematics. Research has shown that the typical student does not in fact lack these abilities [113], so it makes sense to identify aspects of mathematics classes that lead to this erroneous assumption, and to try to eliminate these hindrances. The discussion can be centered around the following possible causes of mathematics anxiety, and their remedies. The meeting leaders or instructors should be encouraged to supply other causes and remedies as well.

- Problem: Math anxiety can be exacerbated by timed evaluations.
  - Solution: Allow plenty of time on evaluations, so that student knowledge and not student quickness is assessed.
  - Solution: Use untimed or take-home exams.
- Problem: Math anxiety can be exacerbated by a fear of forgetting.
  - Solution: Allow students to use calculators on evaluations.
  - Solution: Allow students to prepare and use a "cheat sheet" on evaluations.
  - Solution: Allow students to "purchase" needed formulas or hints during exams for the price of a small deduction in points.
- Problem: Math anxiety can be exacerbated by competitive classrooms.
  - Solution: Grade on a fixed criterion rather than on a curve.

- – Solution: During class allow all students an opportunity to think before calling on a student to answer, and actively involve all students.
- – Solution: Use cooperative learning in class.

- • Problem: Math anxiety can be exacerbated by a fear of word problems.
  - – Solution: Promote familiarity by requiring students to do many word problems.
  - – Solution: Use problems with contexts that are familiar to the students, so that there is no additional anxiety caused by lack of familiarity.
  - – Solution: Teach problem solving strategies to the students.

- • Problem: Math anxiety can be exacerbated by an unwillingness to flounder or a belief that floundering is not part of doing mathematics.
  - – Solution: Make the classroom climate one where it is safe to make a mistake.
  - – Solution: Convey verbally and non-verbally that it is OK (and even expected) for people to struggle with mathematics problems—even mathematicians!
  - – Solution: Encourage and reward perseverance.

- • Problem: Math anxiety can be exacerbated by the fact that mathematics is often a weed-out course for other disciplines.
  - – Solution: Believe that all of your students can succeed, and convey this belief through words and actions.
  - – Solution: Grade on a fixed criterion rather than on a curve.

Finally, we can summarize that mathematics instructors can begin to alleviate their students' math anxiety by being available and approachable, by encouraging their students, by guarding against harmful classroom mythology (such as "only certain people can do mathematics"), by never belittling their students (even in a jocular manner), and by respecting their students' struggles and efforts to master the material.

:40–:45 Provide some time for instructors to reflect on what it means to them to motivate their students, and which of the motivational strategies presented by the experienced instructors or in the handout could work well with their preferred teaching style and with the motivational goals that they have set for themselves. It will be very helpful for meeting participants to have a record of their ideas, so participants can be encouraged to write down their thoughts. A few guiding questions that you could give the instructors to focus their thoughts are

1. What do I think is the outward behavior of a motivated student?
2. What are the most motivating (least motivating) lessons that I have taught, and why were they so?
3. What are my motivational goals for my students?

If there is time, you could have instructors who wish to share their intentions for motivating their classes report their thoughts and plans on what it means to motivate their class, and how they plan to do this.

:45–:50 Wrap up the meeting with a word of thanks, and (as appropriate) time for participants' questions, a summary, suggestions for additional readings, or a questionnaire.

13.6   Meeting Materials

1. Copies of the handout entitled "Ideas for Motivating Students."

Ideas for Motivating Students [1]

1. Intrinsic versus extrinsic motivation

When instructors discuss "student motivation," many end up talking at cross–purposes. Some instructors feel that the secret to "motivating students" is to make class "fun" by handing out candy and taking their class on field trips. Other instructors feel that the secret to "motivating students" lies in finding the correct way to present the mathematics in order to get students excited and eager to learn. Some feel that "motivation" refers to placing the current topic in the correct relationship to other topics studied or to providing "real-world" applications of the mathematics. Finally, other instructors believe that some students just have "motivation," whereas others just don't.

To bring some common ground to discussions of "student motivation," it is helpful to distinguish between *intrinsic* and *extrinsic* motivational strategies.

(a) Intrinsic motivation

    i. Definition of a student who is intrinsically motivated

        Students are *intrinsically motivated* to engage in a learning activity if they recognize that by experiencing the activity they will satisfy a need. Intrinsically motivated students value engagement as *directly* beneficial. The learning activity itself is perceived to be valuable. [21]

    ii. Statements of intrinsic motivation

      • Mathematics interests me.

      • Learning mathematics enables me to think clearly.

      • I feel good when I succeed in class.

    iii. Advantages of intrinsic motivation

      • It can be long-lasting and self-sustaining.

      • Instructors find it satisfying.

      • The efforts to motivate are simultaneously efforts to educate.

      • The focus is on the mathematics rather than rewards or punishments for actions.

    iv. Challenges of intrinsic motivation

      • It can be slow to significantly affect behavior.

      • It can require special and lengthy preparation.

      • Students are individuals, so multiple approaches may be required.

      • It can create student dependence upon the instructor.

      • It can be hard to transport ideas from course to course.

      • The instructor may have to overcome negative dispositions towards the mathematics.

      • The instructor needs to know what interests the students.

(b) Extrinsic Motivation

    i. Definition of a student who is extrinsically motivated

        Students are *extrinsically motivated* to engage in learning activities not because they recognize value in experiencing the activity, but because they desire to receive rewards that have been artificially associated with engagement, or they want to avoid consequences artificially imposed on those who are off-task. [21]

    ii. Statements of extrinsic motivation

      • I need a B- in Calculus to get into the Business School.

      • If I flunk Calculus, I will lose my scholarship.

      • Our instructor will bring us doughnuts if we do well on today's quiz.

---

[1]Adapted from [21], [97], [29], [7], [50], [77], [76], [80], [60], and [96].

    iii. Advantages of extrinsic motivation

- It more readily produces behavior changes.
- It is portable from course to course.
- It involves little effort or preparation for the instructor.
- It does not require extensive knowledge of the students.

    iv. Challenges of extrinsic motivation

- It can distract from the mathematics.
- It can be difficult to devise appropriate rewards and punishments.
- There can be a need to escalate rewards and punishments.
- It does not produce long-term motivation.

(c) The interaction of intrinsic and extrinsic motivation

- Both intrinsic and extrinsic motivation will always be present. The question is, how strong is each in a given situation?
- Extrinsic rewards have been shown to have negative impact on intrinsic satisfaction.
- Accept extrinsic motivation as inherent and necessary to the college classroom, but de-emphasize it. Encourage intrinsic motivation in your students.
- Avoid the conditions under which extrinsic motivation is most damaging to intrinsic motivation:
  - when students would have otherwise been willing to engage in the activities, and
  - when students see the rewards as a means through which to manipulate their behavior.

The following sections will outline several strategies, both intrinsic and extrinsic, for motivating students.

2. Grades (extrinsic)

The majority of students feel that their final grade in a course is extremely important. Sometimes, students report that the only reason they are taking a class is to get a good grade. This can be especially true of mathematics courses that are required for entrance into other programs, such as business or pre-medical. An instructor can harness the students' dedication to achieving a good grade. Some considerations for instructors planning to seriously and deliberately utilize the extrinsic motivation of grades include the following.

(a) View tests and assignments as feedback rather than as a means of teacher control.

(b) Grade higher-order learning outcomes; test for synthesis and evaluation of concepts rather than just factual retention and routine computations.

(c) Consider reducing anxiety by using shorter, untimed, or take-home exams.

(d) Use the weakest extrinsic motivators possible (e.g., regarding attendance or late papers).

(e) Use vocabulary consistent with the interpretation of grades as feedback: "See how well the students are doing," rather than "See if they are performing as they should."

3. Subject matter (intrinsic)

Students frequently complain that they find their classes boring or uninteresting; however, students do not often say that the mathematics itself is inherently boring, perhaps with the exception of required or compulsory classes. Often, it appears that students lack the insight or determination to recognize the interesting aspects of the mathematics for themselves. Most instructors feel that devoting some effort to alerting students to the inherently interesting aspects of the mathematics is worthwhile. Some ideas for revealing the inherent appeal of the mathematics are offered below.

(a) Teach as if mathematics is interesting, informative, and fun (it is!); model enthusiasm.

(b) Expect the students to be interested.

(c) Use *cue words*. The instructor uses cues to show the students where to "look" in order to "see" the appeal of the subject matter. Some possibilities, adapted from [97], are listed in the table below.

Appeal	Examples of Cue Words/Phrases
Novelty	"I think that this is really neat—I haven't seen anything quite the same."
Utility	"This next topic is something that we'll use again and again. It contains really valuable ideas that we'll use throughout the later sections of the course."
Applicability	"As you work through the next section, I think that you'll be pleasantly surprised how relevant it is."
Anticipation	"As you read through, ask yourself what this section of work is hinting at as the next logical step."
Surprise	"We've used 'X' in a lot of different ways. If you thought you'd seen them all, just wait for the next assignment."
Challenge	"Who's up for a challenge? I think that you'll find the next piece of work very interesting."
Feedback	"When you try this, you'll find out whether you really understood yesterday's lesson."
Closure	"A lot of you have asked me about 'X.' Well, finally we're going to find out why that's so."

(d) When designing lesson plans, deliberately plan to reveal intrinsic appeal of mathematics.

(e) Vary the learning and teaching styles that you use in your course.

(f) Tap into the students' natural curiosity.

- Psychological studies have shown that people respond best to stimuli that are different, but not *too* different, from what they are used to.
- Use unexpectedness or surprise.
- Allow them to discover mathematics and make conjectures where it is appropriate.

(g) Emphasize the usefulness, effectiveness, or applicability of mathematics.

4. **Excellence (intrinsic)**

Many people strive for excellence in their lives, and gain a deep sense of satisfaction from a job extremely well done. The pursuit of excellence is one of the most powerful, but least used, motivational tools. Some of the advantages that can accrue from encouraging students in their pursuit of excellence include the following.

(a) Students can receive pleasure from doing something well. To engender this point of view

- help the students create a sense of standards in mathematics, and
- link standards to ability and effort.

(b) Responsibility for learning can be passed to the students.

Ideas for encouraging students in their pursuit of excellence include the following.

(a) Show enthusiasm for excellent work; let students know when they have done particularly well.

(b) Create a classroom climate where students do not feel peer pressure to conform to a mediocre norm.

(c) Protect students who strive for excellence from the criticism and negativity of others.

(d) Encourage feelings of control over outcomes.

(e) For students who have a period of failure, emphasize the unstable nature of grades and performance. Those who tend to believe they will improve tend to remain in the course and do so improve.

(f) Maintain your standards for excellence; give students something towards which to strive.

5. **Success (intrinsic)**

People intuitively calculate their probability of success in a given situation. If students feel that there is very little chance that they will be able to succeed at their studies, then they are often resistant to trying at all. This can be especially true in a mathematics course where a significant portion of the students may doubt their own abilities in the subject. So-called "success models" of schooling are controversial and have been widely criticized. Usually, the basis for criticism is that success is *all* that is emphasized, at the exclusion of appropriate standards and the rigors of the subject matter. Students report, however, that easy success is not satisfying, and not very motivating. The trick with using a motivational strategy based on success is to develop activities at which students can succeed, yet which are substantial enough that students learn much from them. Some ideas for implementing this kind of motivational strategy include the following.

(a) Monitor the difficulty of tests and assignments; try to write test questions that have a moderate risk of failure for a student who is making good progress towards achieving the goals of your course.

(b) Develop and communicate positive expectations. Studies show individual attention affects retention in mathematics and the sciences, especially for female and minority students.

(c) Give differentiated feedback. Point out where a student has done well as well as where a student has done poorly. Give encouraging information about future outcomes.

(d) Attribute success to causes under the students' control.

(e) Use positive vocabulary in every situation: "You are so capable that you should have done better," rather than "Why did you do so poorly?"

6. **Goal setting (intrinsic)**

Some people find that if they do not have a clearly defined idea of what they are trying to achieve, they are less inclined to undertake work. The idea of goal-setting is to develop these clear ideas *before* undertaking the work, and to periodically return to the statement of goals *during* the work. Some recommendations for effectively using goals as a motivational tool include the following.

(a) Emphasize reaching positive goals rather than avoiding negative outcomes.

(b) Allow students to develop their own goals.

(c) Goals should be behaviorally specific: "Study three hours every evening," rather than "Try harder."

(d) Goals should be challenging, yet achievable with effort.

(e) Help students develop strategies for achieving their goals, especially when study skills and other life skills are deficient.

(f) Encourage students to review their goals periodically, and consider whether their current pattern of behavior is helping them to achieve those goals.

13.7 Suggested Reading

- James Cangelosi. *"Classroom Management Strategies."* [21]

  This textbook is designed to help preservice and inservice school teachers effectively manage student behavior. Despite the fact that the students described in this book are school-age children, a great deal of the content is also relevant to the college mathematics classroom.

  Chapter 7 describes the differences between intrinsic and extrinsic student motivation, and suggests ideas for helping teachers to develop classroom activities through which students will be intrinsically motivated to engage with course material. Experienced teachers will recognize this as an ambitious goal for the chapter. The main point is to get students involved in a program of problem-solving, and to develop problems that students feel a need to solve. Some examples are provided, but they are examples of school-level activities.

- Carl Rinne. *"Excellent Classroom Management."* [97]

  This is also a textbook for preservice and inservice school teachers. Again, although the specific examples offered in the book are not all immediately and directly applicable to the college mathematics classroom, many of the theories and techniques described in this book are.

  Chapter 3 ("Motivating Students with Content") describes the concept of intrinsic motivational appeals, and gives examples of ten such appeals. These include: novelty, security, completion, application, anticipation, surprise, challenge, feedback, identification, and competition.

  Chapter 4 ("Designing Content") describes ways to structure classroom activities to "reveal intrinsic appeals." The main suggestion relevant to mathematics is to bring "real-world" problems into your classroom.

- Leo Lambert, Stacey Lane Tice, and Patricia Featherstone, eds. *"University Teaching. A Guide for Graduate Students."* [71]

  Chapter 10 is devoted to "motivating students." A theoretical model for student motivation is presented, which is based on research into human motivation in the workplace. Basically, this model suggests that in order to get motivated, students need to pay attention, they need to see their studies as relevant to whatever they are interested in, they need to feel confident in their ability to complete whatever it is they need to get motivated to do, and they must feel satisfied when they actually do it. Some suggestions for instructional practice are offered, and these are summarized in a table on page 105.

- Wilbert McKeachie. *"Teaching Tips."* [81]

  Chapter 31 addresses student motivation. Again, a model of motivation along the lines of "Students will learn what they want to learn, and have great difficulty learning things that don't interest them" is presented. Unlike the other references, which primarily focus on what instructors can do to "motivate students," the focus of this chapter is on helping students to develop the ability to motivate themselves. The main suggestions for this are (1) to encourage students to formulate goals and (2) to encourage students to recognize and record their progress towards these goals.

- Arthur Schwartz. "Axing Math Anxiety." [102]

  Schwartz gives suggestions for helping students become confident and successful at learning mathematics. He highlights the importance of attitudes, suggesting that students need to view mathematics as critical thinking and problem solving rather than as memorization. He advocates going back to the basics when necessary; this includes reviewing mathematical content as well as study skills, such as how to read a mathematics textbook. He recommends that instructors assist students in obtaining the vocabulary of mathematics by bridging the gap between mathematics terminology and the language of the students. Finally, he points out the importance of support systems.

- Carol Jackson and John Leffingwell. "The Role of Instructors in Creating Math Anxiety in Students from Kindergarten through College." [62]

The authors summarize a survey that investigated the roles instructors play in creating math anxiety in students. The authors found that stressful mathematics experiences predominantly occurred around grades 3 and 4, grades 9–11, or during the freshman year of college. The aspects of instructor behavior that had a negative impact on student attitudes and achievement during college included communication and language barriers, insensitive and uncaring attitude on the part of the instructor, poor quality of instruction, unrealistic examinations, the instructor's dislike for lower-level courses, gender bias, and age discrimination towards nontraditional students. The authors also categorized common instructor behaviors that produced anxiety responses in students, regardless of age, as either overt or covert behaviors. Finally, the authors offer some advice for instructors, such as conveying enthusiasm for mathematics, offering additional reinforcement, and cultivating a classroom climate of respect.

- Jean Jones, *et al.* "Offer Them a Carrot: Linking Assessment and Motivation in Developmental Mathematics." [64]

The authors report on a study evaluating the effectiveness of integrating motivational techniques and assessment practices into a college-level developmental math course. They find that incorporating such techniques did enhance student success. The authors highlight the importance of motivational techniques by describing the relationship between math anxiety and course completion rates.

Chapter 14

Dealing With Difficult Instructor-Student Situations

"The Torch or the Firehose?" is the Math and Physics recitation guide published by MIT and reproduced in [23], on page 185. It notes that,

> The classroom and section meeting are the business end of education. Here students and staff deal together with the stuff of the intellect. Here also idiosyncrasies in both become apparent.

In our experience, and as reported in a survey of graduate student instructors' experiences, [121], most instructors who teach a significant number of students will eventually find themselves in an "idiosyncratic" situation.

We distinguish this meeting from the meeting titled, "Establishing and Maintaining Control in Your Classroom" with the observation that in this meeting we are interested in discussing encounters, or more likely a series of encounters, between instructors and particular students. The earlier meeting can be understood as a meeting about gaining and maintaining students' cooperation with the instructor's plans for the class and how those plans are to be executed. In some sense, that meeting was about overcoming difficulties that may be thought of as in some way inherent to the structure and opportunity that an interactive classroom offers to students. That meeting was about addressing difficulties with an entire class of students. The present meeting focuses on addressing difficulties with particular students. Naturally, there may be relationships between the two phenomena here. For example, a particularly difficult student may cause classroom-wide problems when given access to the opportunities for self-directed behavior that the interactive classroom offers. On the other hand, it is conceptually helpful to distinguish between these two different categories of difficulties—difficulties between the instructor and individual students, and difficulties that may be seen as arising from the nature of the interactive classroom—to the extent to which this is possible.

As in the earlier meeting on "Control in the Classroom," we distinguish between four categories of difficulties: (1) implicit, mismatched ideas of course expectations between instructor and students, (2) difficulties that arise from the nature of the interactive classroom, (3) difficulties between individual students and the administrative structure of the course, and (4) highly interpersonal situations that arise between individual students and the instructor. The principal focus of this meeting is on helping instructors to cope with or find effective ways to deal with "idiosyncratic" situations that are best described by the fifth category.

Because of the highly individual nature of these instructor-student situations, there is very little prescriptive advice that can be offered—and there are certainly no panaceas. The interactive classroom—and many of the relationships that form between individual members of that small community—is complicated. Our focus on providing materials for this meeting has been on (1) creating some scenarios that may help instructors to open up and discuss some of their experiences, and (2) describing characteristics of clear communication. The idea behind including materials on clear communication is to help instructors to communicate more frankly and less ambiguously with students.

As a facilitator for this kind of meeting, it is important to be aware of the highly emotional issues that may be discussed. It is also important to be mindful of some of the practical matters involved. Lastly, there are some points concerning instructor and student beliefs, perceptions and attitudes that are particularly

169

relevant for discussions about instructor-student interactions in elementary mathematics courses. These include the following.

1. In many institutions, there is a perception that elementary mathematics courses are "weed-out" courses designed to screen potential applicants for professional schools. Such a climate, even if it is only a perceived climate, can breed "cut-throat" levels of competition, heightening whatever anxieties students may already possess to the point where it is almost impossible for instructor and students to interact normally.

2. The beliefs that some students hold about the nature of mathematics and how mathematics is learned, have been described in [53] and [101]. The beliefs described in these articles include: mathematics mainly involves memorization of facts, all problems can be solved using the methods in the text, all problems can be solved in five minutes or less, etc. These beliefs probably do not resemble the beliefs that instructors hold about mathematics in the least. In our experience, many instructors may not even be aware that what their students believe about mathematics is very different from what they themselves believe about mathematics.

3. Individuals communicate to each other in a variety of ways, with the literal content of verbal communication being only one such way.

4. There exists a fair amount of information describing substantial changes in the demographics and attitudes of students during the 1990s, [6] and [75]. For example, [75] indicates that the median age of students increased during the 1990s, that more students attempt to hold a part-time job while attending college, and that more students are parents than in the past. In very real ways, the personal situations of students may be very different from those that many mathematics instructors experienced while they were undergraduates. However, this may not be something of which instructors are aware.

5. A significant amount of difficult instructor-student interaction appears to center around questions of evaluation and grading, [121]. This is not news to anyone with experience as a college instructor, but it often comes as a complete surprise to inexperienced instructors. We have found it important to remember that most mathematics instructors were, at some point, "good students" themselves. That is, not only did they achieve at high levels in their courses, they also implicitly understood and played by the "rules" of the course. They studied hard for tests, worked at homework until they had completed every question that time allowed, and they understood the grades that they were assigned to be a reflection of this. Unlike experienced instructors, new instructors have probably not learned that some students may try to work around the implicit rules of the course, or simply reject such basic tenets as, "Your grade in this class is a reflection of your performance on homework, quizzes and tests" altogether.

6. As has been noted by some ([47] and [88]) there is reason to believe that students at different points in their academic careers will have very different views of the nature of the material that they are studying, and what it means to be "intellectually involved" with the material. For example, beginning students often view all material as very authoritative and factual. Their expectations for being involved with the material may extend only as far as thinking that they need to memorize this "factual content." The views of the students and the corresponding views of the instructor in this matter may not be well aligned.

In summary, bear in mind when facilitating this meeting that all you will usually hear is one side of the interaction—the instructor's side of the interaction. As facilitator, you will also need to remember that what instructors tell you will be a strong reflection of what individual instructors consider to be fundamental "laws" of the classroom. Be mindful of the fact that it can be quite a shock for instructors to discover that not everybody recognizes or operates according to these "laws." Even if it does not directly affect the way that you facilitate the meeting, it is helpful to bear in mind that much of what you may hear will be based on implicit beliefs that instructors assume everybody shares. Finally, it is important to be receptive to instructors, and to genuinely want what is best for everyone involved. The important point here is that there are more people involved than just the instructor who feels upset or frustrated. Difficult instructor-student interactions can also spill over to affect the quality of the classroom as a learning environment for other students.

14.1   Description and Purpose of the Meeting

This meeting is intended to be a fifty minute session, although depending on the number of participants, and the amount that they have to say, it could be much shorter or much longer. Given the distress that these kinds of issues often provoke among instructors, it may be more comfortable (for certain instructors) to discuss their situation one-on-one with a trusted colleague. The only information distributed at this meeting will consist of some tips on assertiveness and managing confrontations. The main purpose of this meeting is to provide instructors with a forum for getting difficult experiences with students off their chest, for listening to the kinds of challenges that other instructors have faced, and for giving and receiving support. This experience is intended to encourage participants to think about what they hope to achieve in their interactions with students. Participants should be encouraged to think about what kinds of behavior they consider acceptable from students, and how they will respond to unacceptable behavior.

With the understanding that the decision between addressing the behavior of the entire class and addressing the behavior of individual students is difficult, and can have serious consequences, this meeting can also be used as a venue to encourage new instructors to seek out the opinions of their more experienced colleagues before committing to a course of action.

14.2   Goals for the Meeting

1. To provide instructors with an opportunity to describe difficult interactions that they have had with students.

2. To provide instructors with an opportunity to hear about the kinds of situations that their colleagues have had to deal with, and that they may face at some stage, even if they haven't already.

3. To provide instructors with an opportunity to hear others' ideas of how to deal with difficult situations.

4. To provide instructors with an opportunity to think about what is important to them in instructor-student interactions.

5. To provide instructors with an appreciation for some techniques for handling difficult people, for being appropriately assertive, and for managing confrontations.

6. To provide instructors with a forum for hearing others' thoughts on the issues outlined above.

14.3   Preparation for the Meeting

1. Gather institutional or departmental guidelines on instructor-student interactions, if available.

2. Review and modify the handout entitled, "Difficult Student Behaviors." Customize the list to reflect the behavior of the student body at your institution.

3. Review the handout entitled, "Strategies for Clear Communication and for Managing Confrontation."

4. Circulate the agenda for the meeting to all participants well in advance.

5. Make copies of meeting materials, and any questionnaires that you plan to distribute, and confirm attendance with participants.

14.4   Agenda for the Meeting (to be distributed to participants)

Note: As shown in the next section, this meeting can be run in several different ways. The format that the meeting leader selects should be chosen to meet the needs of the meeting participants. The agenda that you choose to distribute should be modified to reflect the kind of meeting that you plan to run, and any special need or perspective that is not reflected in the sample agenda given below.

1. A discussion of the differences between dealing with difficult students and difficult teaching situations.

2. A description of some general strategies for clear communication, and some general strategies for managing disagreements and confrontations.

3. A discussion of the different kinds of difficult behavior that students often exhibit, and strategies for dealing with specific kinds of behavior.

4. Opportunities to speak with other instructors about difficult student behaviors.

14.5   Outline for Running the Meeting

As with all of the meetings in this volume, this meeting can be conducted formally, informally, or semi-formally depending on the meeting leader's intentions, and the needs of the meeting participants. To help you to choose from these options, all three approaches are given in detail.

1. **Formal Meeting:** This meeting format would be appropriate if the meeting leader wishes to impart a large amount of information, such as strategies for dealing with different situations and difficult behaviors that are commonly encountered. This kind of meeting format is also appropriate if the participants have expressed interest in the topic, but are not comfortable speaking about their own experiences. A detailed schedule for a formal meeting is given below.

   :01–:05 Introduce the meeting, and briefly review the agenda. Ensure that the participants appreciate that the focus of this meeting will be dealing with individual students who exhibit difficult behaviors, rather than dealing with class-wide behaviors. Make sure that instructors appreciate that dealing with class-wide problems by singling out individuals is not always the best strategy, and conversely, that dealing with difficult individual behavior by addressing the whole class can be counter-productive as well. It will be helpful if you, or other experienced instructors, are able to provide examples from your own experience.

   :05–:15 Distribute the copies of the handout on "Techniques for Clear Communication." Explain the view that addressing student behavior is always a difficult and emotional issue. Make sure that the participants understand that unclear, ambiguous or vague communication can exacerbate a difficult situation. Ensure that the participants appreciate that, in order to concentrate on the issues they want to address, i.e., student behavior, it is critical to be absolutely clear and unambiguous in their communication. You could run through what you see as being the most important issues of clear communication indicated in the handout. Putting these highlights on an overhead, or writing them on a chalkboard as you go can add emphasis. Make sure that participants have the opportunity to ask questions, and that the questions are answered. Experienced instructors who attend the meeting can be a very valuable resource for answering participants' questions. Call on them to suggest answers as often as you feel appropriate.

   :15–:25 Distribute copies of the handout on "Managing Conflict." Give the meeting participants a few minutes to read through the sections. Explain that even when the instructor is careful, sensitive and well-intentioned, students can take any attempt to address behavior very negatively, which can lead to an unpleasant confrontation. Try to get across to participants that confrontation is not a kind of failure on their part. In a sense, by creating a confrontation, the student has once again demonstrated a capacity for difficult behavior. Make sure that participants realize that confrontation is something to be dealt with, and worked through as calmly as possible; it is not something that needs to be feared, or avoided at all costs. Make sure that the participants appreciate that it is often more helpful, at least in the long run, to really work out a problem, rather than to pretend that the problem doesn't exist.

   :25–:45 Distribute copies of the handout entitled, "Potentially Difficult Instructor-Student Interactions." Give the participants a chance to look at the handout, and reflect on the questions. In addition, participants could give examples from their experiences of difficult student behavior. Alternatively, the meeting leader can ask participants to pair up, and describe to each other any instances of difficult student behavior that they have encountered, and what they did about it. You could ask the pairs to think about the situations described on the handout, and to generate suggestions for dealing with these behaviors. After some discussion, you could have pairs report to the whole group, and have them describe how they think it would be appropriate to handle a given situation.

:45–:50 Wrap up the meeting with a word of thanks, and (as appropriate) time for participants' questions, a summary, suggestions for additional readings, or a questionnaire. This is a good opportunity to point out that dealing with individuals' behavior and class-wide behavior can be intertwined, but can also be quite different issues. Point out that it is usually a good idea to consult with a trusted, experienced colleague before taking any definitive or irreversible actions. It can also be a very useful idea to document, immediately, any situations that occur. It is not unheard of for students to complain to senior figures in the department and university administration. Instructors need to have a clear picture of what actually transpired should such unpleasant circumstances arise.

2. **Informal Get-together:** This meeting format would be appropriate if the meeting leader was mostly concerned with having instructors speak with each other about their experiences, and to come up with ideas for how to deal with specific examples of difficult student behavior. As this suggests, the informal get-together would also be appropriate if many meeting participants had expressed the desire to talk about their own personal experiences. In such a meeting, a detailed schedule is less useful, as the meeting will be driven by the participants' needs and interests, rather than by any particular person's ideas of what should be covered. Handouts can be used as a supplement to the ideas exchanged in the meeting, and as a site for generating ideas, rather than the focus of the discussion. A list of possible discussion questions or topics is given below. The meeting leader could use these to sustain the get-together, or to spark the conversation. The handout entitled, "Potentially Difficult Instructor-Student Interactions," could also be employed to begin the discussion, or a similar document could be prepared in advance by the meeting leader and distributed to participants at the beginning of the meeting. Participants would be asked to read through the situations described on the sheet, and to find any that resemble incidents that they have experienced personally. Once participants have read and thought about the situations described, the meeting leader could ask them to turn to a neighbor and describe which of the situations described on the sheet particularly caught their attention, and why.

Possible Questions for Discussion

(a) Are there differences between dealing with difficult students and dealing with difficult teaching situations?

(b) Addressing difficult behavior is often upsetting for instructors and students. What can be done to manage disagreements and confrontations?

(c) What kinds of difficult behavior *do* students exhibit? What strategies can be employed to deal with different kinds of behavior?

(d) What difficult student behaviors have you had to deal with?

If a more structured setting is preferred, the meeting leader could make an agenda and circulate this to participants before the meeting. This may help to encourage the amount and quality of participation in the meeting by giving participants a chance to think before coming to the meeting. A sample agenda is given below.

Sample Agenda for Informal Meeting

Item 1. Discussion with others of how you see the students in your class, and the kinds of behaviors they exhibit. A chance to describe what you think the relationship between the actions of individuals and the behavior of the whole class may be. A chance to describe to others what you feel you would like to see happen differently, on the level of the individual student and on the level of the class as a whole. An opportunity to talk with others about what the possible outcomes of your actions could be. Everybody present will have a chance to talk about the particular issues that they are dealing with, and receive support, ideas and advice from other participants.

Item 2. What can be done now? What possibilities does the current situation present? There will be opportunities for participants to express what they would like to change about the behavior that individuals exhibit in their class. Participants will also be able to receive advice and ideas from other participants and experienced instructors to assist them in sorting out particular situations. This could be a good time to distribute the handouts on strategies for clear communication, and strategies for managing confrontation, in support of the participants' discussion.

Item 3. Discussion of some difficult student behaviors that people have experienced and tried to deal with in the past, and the results that they have seen. This could be a good time to distribute the handout entitled, "Potentially Difficult Instructor-Student Interactions." It could also be a good time to indicate some of the behaviors on the handout that you have had to deal with in the past.

Item 4. What can we learn from this? How can we conduct ourselves so that we don't have to go through this kind of an experience again and again and again?

3. **Semi-formal Meeting:** This is a compromise between the first two approaches, and could be used if the meeting leader felt strongly that participants ought to discuss the content of the handouts in the meeting, whereas many meeting participants had expressed interest in talking with other instructors about difficult student behavior. This format is also helpful if the meeting leader wants participants to have the opportunity to talk with each other about their experiences, but also feels that it is important for participants to be exposed to a more theoretical or abstract description of dealing with difficult behavior.

This kind of meeting is more difficult to lead than the first two, as it requires the meeting leader to pay close attention to the time, and to make impromptu decisions about how long to let things go on. This requires the meeting leader to have an acute sense of how much the participants are getting out of each segment of the meeting, and whether the meeting time may be more profitably spent doing something else. This also requires the meeting leader to recognize when items on the agenda are less important, and retain enough flexibility to omit less important agenda items.

One idea for running this kind of meeting is to make the meeting participants active agents in deciding which agenda items to emphasize, and which to omit. This can be done by preparing a loose schedule for the meeting with suggested amounts of time for each item. This would be copied and distributed to participants as they arrive at the meeting. As the meeting develops, if it becomes clear that there will not be sufficient time to treat every item remaining on the timetable, the meeting leader can ask the participants to decide which of the remaining items they would like to focus on in the remaining time. If the meeting is run with this kind of strategy, the meeting leader should be careful to organize the agenda so that items of interest to everyone appear early in the agenda, and items with less general appeal are scheduled for later. If the meeting leader does not have a clear sense of how popular or imperative each agenda item is likely to be, a list of possible agenda items could be circulated prior to the meeting, and participants asked to indicate which of the possible items they would be most interested in discussing.

A suggested timetable is given below.

(About 5 minutes.) We will discuss the relationship between the behavior of individual students and the behavior and atmosphere of the entire class. The discussion will also touch on the appropriateness of dealing with individual behavior in a way that affects the whole class, and the effects of dealing with class-wide behavior by singling out individuals.

(About 10 minutes.) We will discuss the need for clear and unambiguous communication when dealing with students who exhibit difficult behavior. This discussion will touch on the fact that dealing with students' behavior is an emotional issue, both for the instructor and the student, and the subsequent need for very clear communication on exactly what the problem is, and what can be done to address it.

(**5 to 10 minutes.**) We will consider our experiences of difficult student behaviors, what we tried, what the outcomes were, and what we may do differently in the future.

(**5 to 10 minutes.**) You will receive a description of some of the skills that help to manage tense or confrontational situations, and of experiences that people have had in dealing with these kind of situations.

(**About 10 minutes.**) You will receive a description of some of the categories of difficult behavior that are frequently encountered.

(**About 5 minutes.**) You will have an opportunity to network with some other instructors, so that you have people to speak with when you have to deal with difficult behavior. We'll also try to recognize that all of us have to deal with difficult behavior, and that this is an emotionally and intellectually difficult, but essential, part of teaching.

14.6 Idea for Expediting the Meeting

Instead of organizing a "full-blown" meeting, distribute the handouts and then invite interested instructors to an informal meeting. To achieve this, circulate an invitation asking people to attend if they feel that they would like to. The tone should be inviting, and very deliberately non-judgmental. A sample invitation is given below.

Colleagues,

In response to inquiries from interested instructors, we have planned a meeting to be held at <time> on <day and date>. We hope that this meeting will serve as a kind of informal get-together where we are free to discuss and listen to each others' experiences in dealing with difficult students.

If you have ideas, or if, like me, you would like to talk about experiences in your class in an informal and relaxed atmosphere, please consider yourself to be very warmly invited to attend.

Sincerely,

<meeting leader>.

14.7 Meeting Materials

1. Copies of institutional or departmental guidelines or regulations covering instructor-student interactions.

2. Copies of your modified version of the handout entitled "Potentially Difficult Instructor-Student Interactions."

3. Copies of the handout entitled "Techniques for Clear Communication."

4. Copies of the handout entitled "Managing Conflict."

5. (If you are planning a semi-formal meeting.) Copies of your timetable for the meeting.

Potentially Difficult
Instructor-Student Interactions

1. Kenny approaches you after a major test, obviously agitated. He says that although he feels that he "really understands" the material he had no idea of how to approach many of the problems on the test.

2. Liz is doing very well in the class, and completes every homework assignment on time. Her scores are usually the highest in the class. However, whenever you include a problem that is not "just like an example from class" she struggles.

3. Larry is alert and attentive in class, and is often the first to respond to your questions. Even when you direct your questions to other students in the class, Larry often speaks up instead.

4. Due to low turnout in class, you announce a policy of weekly quizzes. The students who are present in class grumble a bit, but don't seem too upset. Initially, you plan to have the quizzes at the beginning of class. Robby, who normally doesn't come to class faithfully, appears and takes the quiz. When you collect the quizzes, he gets up and, in clear view of everyone, begins to walk out of the room.

5. Despite low scores on every test, Daryl insists that he knows the material in depth—after all, he had most of it in "AP" last year. Daryl does very well on problems that involve manipulating quantities, especially numerical quantities. Daryl comes to you one day and asks that you spend less class time discussing applications and interpretations of mathematical problems, and more time going over the types of problems that will be on the exam.

6. When you hand back papers, there always seem to be students who are disappointed with their scores. John is absent from class that day, and comes to pick up his test from your office. When he sees the score, he becomes obviously upset, and starts flipping through the pages in an agitated way. He comes to the end of the paper, and says, "I don't understand. I worked my butt off studying for this test."

7. Melinda asks to meet with you after the midterm. Her score was 72/100, the class mean was 69/100, and the class normally has a C^+ average. Melinda asks if there is any way that she can do extra work for extra credit, as there is no way she can afford to get a 'C' in this class.

8. A few weeks into the semester, you get an E-mail from Eric. Eric was in your section last semester and earned some kind of a 'C' in the course (in fact it's probably generous to say he 'earned' a 'C'). Eric says that he heard that it was possible to get the grade changed up to six months after the course, and that he just heard that he needs a 'B' to get into business school. In his message he asks you to let him know what he has to do to get the grade changed, and confidently states that he will be happy to "... do whatever (if anything) is required to rectify this situation. Thank you."

9. Despite obvious efforts, Nancy is having a hard time coming to grips with the math that you have been explaining in class. Nancy regularly attends office hours, and makes appointments to see you outside of these, too. Nancy seems earnest in her desire to learn the material for the course, and she is very appreciative of your efforts, but these meetings are seriously eating into your schedule.

Techniques for Clear Communication

1. Active Listening

It is important to draw a distinction between *hearing* and *listening*. In [112] the authors define hearing to be a purely physical activity in which acoustic energy in the form of sound waves is changed to mechanical and electrochemical energy that the brain can understand, while they define listening to be the psychological processes that allow you to attach meaning to the patterns of energy heard.

Keys to Effective Listening	A poor listener tends to...	A good listener tries to...
1. Judge Content, Not Delivery.	lose interest if the delivery is poor.	identify the main points and ignore slips and gaffs.
2. Hold Your Fire.	argue with speaker when the opportunity arises.	wait until the speaker is finished before deciding whether understanding is complete.
3. Listen for Ideas.	home in on facts.	pick out the theme.
4. Be Flexible.	record almost everything said.	use a note-taking intensity appropriate to the setting and content of the speech.
5. Work at Listening.	show no energy, no attention, or tries to fake it.	give signals when comprehending and look at the speaker.
6. Resist Distractions.	respond to every distraction that comes along.	ignore distractions and the bad habits of the speaker.
7. Exercise Your Mind.	give up easily when confronted with intellectual substance and use phrases that put all responsibility for understanding on speaker.	enjoy the challenge of heavy material.
8. Keep Your Mind Open.	react (or overreact) to specific words.	avoid getting upset over specific words and place words in context.
9. Capitalize on the Fact that *Thinking* is faster than *Speaking*.	daydream or fidget with slow speakers.	weigh evidence, strive for understanding, listen to tone of voice and read body language.

In [112] the authors present characteristics of good and bad listeners. The table given above is adapted to circumstances where an instructor is listening to students.

In [25] the author suggests ways to improve listening skills. The list given below is adapted for instructors trying to become more effective at listening to students.

(a) You must care enough to want to improve. Trying to improve without really wanting to, or thinking that you need to improve, requires a supreme effort.

(b) Find or create an area for conversation that will be free from interruptions. A quiet place will help you to focus on what the student is saying to you, rather than on a myriad of interruptions.

(c) Try not to finish the student's sentences.

(d) Pay careful attention to what is being said. Don't stop listening because you want to rebut a particular point. Instead of butting in, take notes on what the student has said incorrectly while the student is speaking. When the student has finished speaking, address each of the misconceptions one at a time.

(e) Be aware of emotionally charged, 'red flag' words or phrases when they are used in the conversation. 'Red flag' words or phrases can indicate that the speaker is becoming angry or agitated. You may make more progress by taking a break and allowing some time for everybody to calm down.

2. **Speaking**

   While listening is about understanding clearly what *others have to say*, speaking involves expressing what *you have to say* clearly so that your listener understands what *you have to say*.

   In [25] Cava lists three main reasons why people feel they are not good speakers. These reasons have been adapted to instructor-student exchanges and listed below.

   (a) **The speaker may have trouble choosing the right words.** Sometimes, having heard what the student has to say, the instructor will know *how* they want to respond, but won't be able to find quite the right words. The following are a few options for dealing with this.

   - Suggest a break where the student leaves for a few minutes.
   - Say that you understand the student's position, but need some time to think it through. Either schedule a follow-up meeting, or write a response for the student.
   - Consult with other instructors who have the reputation of being able to handle difficult student situations.

   (b) **The speaker thinks that the student can't follow what's going on.** It may be necessary to ask the student for confirmation that they have understood.

   (c) **The instructor or student may be a "motor mouth."** In this case, the speaker simply has difficulty keeping the conversation short, sweet and to the point. If the problem is a student with a "motor mouth," then the instructor can ask guiding questions, such as those below.

   - "That's very interesting, but can you explain more thoroughly why this is important to talking out of turn in class?"
   - "Yes, I agree that it's important for everyone to have a chance to participate in class. How does that connect with what you're saying now, though?"
   - "I'm sure that your visit to New Jersey will be very interesting, but let's get back on track, here. You say that you shouldn't be penalized for those tests you missed because... ?"

   If the problem is an instructor with "motor mouth," the instructor can try the following.

   - Using note or outline form, write out exactly what you want to discuss with the student. Keep your notes handy during the encounter with the student, and stick to the points that you have written down.
   - If "motor-mouth" is a general problem in other areas of your personal or professional life, you can try rehearsing by speaking into a tape recorder. Play the recording back, and try to think of ways to express yourself more succinctly.

3. **Paraphrasing**

   In [25], Cava describes the skill of paraphrasing. Part of her introduction is reproduced below.

   > Paraphrasing means: to express meaning in other words; to rephrase; to give a message in another form; to amplify meaning.

   Paraphrasing may be used to confirm that you have understood what the speaker has told you, to clarify your understanding if you don't feel that you understand what the speaker has told you, or to bring out discrepancies in what the speaker is telling you.

4. **Feedback**

   This technique provides a way of structuring interactions with students who have exhibited difficult behaviors. By having some plan for how the interaction will proceed, the instructor can improve his or her chances to clearly communicate with the student without becoming distracted by side issues.

   Feedback is a technique that is very suitable for use when addressing difficult student behaviors. When using the feedback technique, you identify what another person has done that annoys or upsets you, tell the person exactly what it is that annoys or upsets you, and then give the person an opportunity to do something about it. A central idea when using the feedback technique is to deal with situations as quickly as you can.

In [25] Cava gives steps for the feedback process, and guidelines for effective feedback. We have adapted those steps and guidelines to instructor-student encounters. The steps and a list of adapted guidelines is given below.

Steps in the Feedback Process

(a) Describe the problem or situation to the student who is exhibiting the difficult behavior.

(b) Define what feelings or reactions you have towards the behavior.

(c) Suggest a solution, or ask the student to provide a solution.

Guidelines for Giving Effective Feedback

(a) **Be sure the student is ready.** If the student appears distracted, seems very resentful of being taken aside, or keeps asking questions like, "Is this going to take long? I really have to be somewhere else," then it may not be the right time to give your feedback.

(b) **Base your comments on facts, not on emotions.**

(c) **Be as specific as you can.** When you can point to specific instances of the student's behavior that you considered inappropriate, you will have a much more dramatic impact on the student than vague statements like, "It bothers me when you talk in class, because others can't hear."

(d) **Give feedback as soon after the event as possible.** The closer in time that you can give the feedback, the more likely it will be that the student can recall the behavior that you wish to address.

(e) **Try to find a private place to give feedback.** Giving highly critical feedback in front of others can sometimes be more damaging than helpful.

(f) **Concentrate on what the student can actually change.**

(g) **Request the student's cooperation.** You may ask for the student to help make class the best learning environment possible by reconsidering the choices that the student makes about how to behave in the classroom.

(h) **Focus on one thing at a time.** By presenting too much information, or addressing too many student behaviors at once, you may leave the student confused and very upset.

Managing Conflict

1. Avoiding Conflict

Many people automatically attempt to avoid conflict, with the assumption that the path that offers the least resistance in the *short term* always represents the best solution. When dealing with difficult situations, the choices that represent the easiest path in the short term may not always turn out to be the best choices in the long term. Consider the example of a student who demands a chance to do extra work to make up their grade. The short term problem of the belligerent student is easily solved by agreeing to their demands. However, this may create a long term problem when other students demand the same opportunity.

2. Understanding Conflict

When intelligent, creative, determined people work together, differences of opinion and conflicts are almost inevitable. Conflict is not inherently good or bad. In [27] three common causes of conflict are identified.

Adversarial Action: People look at disagreements as "win-lose" situations. In the context of addressing difficult student behavior, if the instructor or the student believe that there is something to be won or lost, for example a sense of personal autonomy or a sense of classroom authority, then it may be more difficult to see the behavior as something that can be changed.

Tightly Held Positions: People see no need to achieve goals together. Instead of trying to work out differences, people harden their position, and limit involvement with others. When addressing difficult student behavior, it is possible that instead of recognizing their behavior as inappropriate, the student will understand that the instructor is trying to say that the behavior is somehow *wrong*. In such a case, the student may not want to admit that they are *wrong*, whereas the instructor simply wants the behavior to stop.

Emotional Involvement: People become attached to their positions, methods and way of working out problems. The possibility of emotional involvement when addressing difficult student behavior is clear. Students may not see their difficult behavior as a problem, and be upset or offended by an instructor's efforts to address the behavior. On the other side, some instructors are not willing to address student behavior until they have become upset by the behavior.

3. Emotional Reactions to Conflict

The following emotional reactions to conflict are adapted from [27].

- The instructor may decide to directly confront the student. Instead of the intended frank and direct message, the student understands that the instructor is being aggressive. In turn, the student may reply in a way that the instructor feels is aggressive, and the situation escalates into an **aggressive conflict**.

- The instructor feels that the best way to address behavior is indirectly. The instructor concocts situations to make the student look bad, or to make others feel that they are the victims of the difficult behavior. The instructor tries to **manipulate** the situation.

- The instructor keeps telling him or herself that it's easiest to just grit their teeth and put up with the behavior, because the semester will soon be over, and they'll never have to see the student again. The instructor continuously **postpones material that causes conflict.**

- Instead of responding to difficult behavior as difficult behavior, the instructor responds as though the behavior is perfectly acceptable. The instructor effectively **gives in.**

4. Preferred Responses to Conflict

- Direct energies towards solving difficulties, rather than using the energy to worry about the difficulties.

- Respond to students in a rational way, even if they are reacting in a very emotional way.

5. **Six Steps for Confronting and Resolving Conflict**

The following set of steps are adapted from [27], and intended to provide a framework for navigating conflicts.

(a) Both the student and the instructor need to acknowledge that conflict exists.

(b) Identify the real reason for the conflict. Both parties must agree that they have found the real reason if they are to make further progress.

(c) Both the instructor and the student need to put forth their point of view. Both parties must listen carefully and *not interrupt* while the other is speaking.

(d) The instructor and the student must explore ways to resolve the difficulty. These potential solutions must be ideas that both the instructor and the student can live with.

(e) The instructor and the student agree on a solution.

(f) The instructor follows up on the solution to make sure that it actually works.

14.8   Suggested Reading

Not very much seems to have been written about difficult instructor-student interactions in mathematics. There are some books and articles in the more general college teaching literature, some in the professional teaching literature for discussion leaders (especially those who work in composition programs) and some in the form of "self-help" books for business people. Some examples are given below.

- Wilbert McKeachie. *"Teaching Tips."* [81]

  This includes a chapter (Chapter 24) describing several types of potentially problematic student behavior, and includes McKeachie's thoughts on how he deals with each. The book also includes a section in Chapter 6 (testing) that suggests some ideas for reducing the "student aggression" that sometimes surrounds evaluation issues, and a section of Chapter 21 (Large Classes) dealing with order and discipline.

- Sharon Biaocco and Jamie DeWaters. *"Successful College Teaching. Problem-Solving Strategies of Distinguished Professors."* [6]

  Chapter 10 of this book includes a list of "teaching difficulties" identified by surveying teaching assistants from a variety of disciplines. The chapter also includes one or two possible solutions offered by distinguished faculty recognized as accomplished teachers. Areas of difficulty addressed are: (1) underprepared students, (2) assessment techniques, (3) cheating, (4) poor attitudes, (5) personal attacks, (6) lack of credibility, (7) student irresponsibility, and (8) grade-grubbing.

- Patricia Shure, Beverly Black and Douglas Shaw. *"The Michigan Calculus Program Instructor Training Materials."* [103]

  This handbook includes a list of potentially difficult instructor-student interactions that could be used to supplement those given here.

- Anne Statham, Laurel Richardson and Judith Cook. *"Gender and University Teaching. A Negotiated Difference."* [108]

  Chapter 4 (Authority Management) describes some of the behaviors that students exhibit, and lists responses that the faculty members recalled using to respond. The information is organized according to academic rank and gender of the faculty member.

- Roberta Cava. *"Difficult People."* [25]

  This is a book for business people. The book contains advice for dealing with a variety of situations that can occur in the workplace. For example, dealing with situations from angry customers to deceitful coworkers. The book also includes descriptions of techniques for improving the clarity of communication, although again, this is situated in a business context.

- Muriel Solomon. *"Working with Difficult People."* [106]

  This book identifies a number of "personality types" in terms of the problematic behavior that they exhibit. Advice is given for how to deal with bosses, coworkers and subordinates who fit each personality type described. Some of the strategies for dealing with difficult subordinates may be helpful for college instructors. The underlying assumptions about the workplace may diminish the utility of some of the advice, at least for instructors. For example, instructors usually cannot do anything that is analogous to "firing an employee."

- Dale Winter and Carolyn Yackel. "Novice Instructors and Student-Centered Instruction: Understanding Perceptions and Responses to Challenges of Classroom Authority." [122]

  This article describes the difficult instructor-student interactions reported by a group of graduate student instructors (GSIs) in a large mathematics department. The article lists five priorities that influenced GSI's perceptions of and responses to students. These priorities were: protecting evaluative authority, following the regulations of the course, protecting rights to make pedagogical decisions, protecting credibility as a teacher, and ensuring a basic level of respect for self and others. The article suggests guidelines for course directors and faculty members to help beginning instructors learn from difficult interactions with students.

Chapter 15

End-of-Semester Administration

By the end of their first semester of teaching, new instructors should have become somewhat comfortable with the day-to-day issues of planning their lessons, meeting students in office hours, grading homework, etc. Unfortunately, the end of the semester brings new and different stresses and struggles. The traditional pressure of grades places added importance to the students on mastering new and old material. In addition it provides temptation for students to cheat during the final exam or to bargain (or beg!) for grade changes after the exam. New instructors may be caught unaware by some of these issues. This meeting is intended to preempt some of these difficulties by conveying institutional guidelines as well as providing practical and moral support for dealing with these issues. It is also a forum for conveying administrative information, such as how to schedule a review session, what is expected of instructors on exam day, how to calculate final grades, and any post-semester responsibilities the instructors may have.

15.1 Description and Purpose of the Meeting

This meeting is intended to be a thirty to fifty minute session. The meeting is intended to be a venue for distributing information, and for communicating to instructors an unequivocal statement of responsibility. It is not necessary for experienced instructors to attend such a meeting, although these instructors are an invaluable resource for the meeting leader. This meeting is primarily targeted at those instructors who are new to teaching in your department.

Note: An optional, additional purpose for this meeting is to provide a demonstration of the process of calculating final grades, according to the methods and conventions of your institution. This will increase the length of the meeting. A handout describing one particular grading scheme is appended. This could serve as a starting point for developing handouts and overhead slides illustrating the process of calculating final grades appropriate to your institution.

15.2 Goals for the Meeting

1. To provide instructors with the precise details of the administrative activity that is expected from them at the end of the semester.

2. To communicate an unequivocal statement of each instructor's responsibilities to the students, to the course, and to the university.

3. To provide instructors with an opportunity to ask questions about the parts of the end-of-semester administrative process.

4. To provide instructors with ideas for dealing with attempted cheating on the final examination.

5. To provide instructors with an awareness of the kinds of grade complaints and disputes that they may encounter, and with ideas for how these situations may be handled.

6. To provide instructors with an understanding of your institution's grading scheme and conventions.

7. To provide instructors with a checklist of information to give to students before they take the final exam.

8. To provide instructors with a forum for hearing others' thoughts on the issues outlined above.

9. (optional) To acquaint instructors with your institution's formula or recipe for assigning final grades.

15.3 Preparation for the Meeting

1. Gather institutional or departmental guidelines on assigning grades, if any such guidelines exist.

2. Modify the handout entitled "Administrative Issues for the End of Semester," to bring it into line with the conventions and practices of your institution.

3. Modify the calendar, highlighting important dates to suit your institution and semester.

4. Create a flow chart detailing the official grade reporting procedure for your institution.

5. Collect copies of the official grade reporting forms for your institution, if they are available.

6. (Optional) If you are planning to demonstrate a method of calculating final grades, then you could prepare a handout or overhead projector slides to accompany this demonstration. An example handout is included in the meeting materials.

7. Circulate the agenda for the meeting to all participants well in advance.

8. Make copies of meeting materials, and any questionnaires that you plan to distribute, and confirm attendance with participants.

15.4 Agenda for the Meeting (to be distributed to participants)

Notes: The agenda suggested below is quite comprehensive, and it may include items that are not relevant to your institution or teaching environment. Because this meeting is intended to help less experienced instructors deal with the examination period confidently, it will pay to consider each item on this agenda very carefully.

1. Before the Final Exam

    (a) Scheduling review sessions
    (b) Information for class (date, time, location of final)
    (c) Alternate exam
    (d) Instructor commitments, travel plans, substitutes

2. Final Exam Day

    (a) Proctoring/grading versus instructor commitments
    (b) Exam room security

3. Cheating

4. Submitting Grades

    (a) Before submitting
    (b) Submitting grades
    (c) People who are leaving town early

   5. Pass-Fail, Incompletes, and Excused Absences

   6. Communicating Grades

      (a) Returning papers

      (b) Methods and precautions

      (c) Communicating grades

      (d) Grades via E-mail

      (e) No posting / no grades through departmental offices

      (f) Privacy

   7. Handling Complaints

      (a) Being ready for complaints

      (b) Checklist for explaining grading decisions

      (c) Common student arguments for special consideration

      (d) Belligerent students

      (e) Improper conduct

   8. Changing Grades

   9. Post-Semester Responsibilities

   10. Determining Final Grades (Optional)

15.5   Outline for Running the Meeting

:01–:05 Introduce the meeting, and briefly review the agenda.

:05–:15 Distribute copies of the handout that you have developed or adapted from the handout entitled "Administrative Issues for the End of the Semester." Distribute copies of institutional guidelines, if they are available. Spell out what each instructor is responsible for doing before, during, and after the final exam. Make an explicit, unequivocal statement of each instructor's responsibility. Avoid vague or uncertain statements. If an instructor asks a question that cannot be answered on the spot, promise to get back to the person with the answers, *and then do!* The instructors should be left with a very clear idea of exactly what is expected of them during the exam period. Reference items in the handout, "Administrative Issues for the End of Semester," as appropriate. Another idea is to make a calendar like the one in the appendix, and then to distribute this to instructors.

:15–:20 Reiterate the content of the meeting in Chapter 10 on cheating and exam proctoring. By the time of the final exam, most of the instructors will have had experience with giving examinations, and there will be no need to overdo this subject. A quick summation of the main points is probably all that is required, although it is worth reminding instructors to be careful about accusing students of cheating. This could be a good opportunity for questions and answers.

:20–:30 Describe the mechanics of the official grade reporting procedure at your institution. If they are available, bring copies of official grade reporting forms to show to the instructors. A diagram, like a flow chart on an overhead, may be a useful visual aid to show the grade reporting procedure. Many instructors may attend the meeting for this information alone. Make sure that there is an opportunity for them to get answers to all of their questions.

:30–:35 Describe any non-standard grades, such as incompletes, excused absences, and pass/fail options, that your institution allows. Specify the circumstances under which students may qualify for these non-standard grades. Try to be as definite and clear as you can here. Instructors who are new to your institution, or to teaching, will not know the conventions of your institutions, and practices vary widely from institution to institution.

:35–:45 Describe the official process for returning exams to the students, if your institution has one, or else describe some of the methods that you are familiar with. As an alternative, if any experienced instructors have decided to attend the meeting, you could have them describe the methods that they use to return papers to students. This is also an opportunity for instructors to share stories of students that they have had to deal with, and what they did. Clearly communicate the message that a thoroughly thought–out grading system applied judiciously to every student is the best insurance against belligerent students.

:45–:50 (If you are planning to include a demonstration of determining final grades, delay this portion of the meeting until after that is completed.) Wrap up the meeting with a word of thanks, and (as appropriate) time for participants' questions, a summary, suggestions for additional readings, or a questionnaire.

:45–:?? (If you are planning to include a demonstration of determining final grades.) Demonstrate the process of calculating final grades using your institution's accepted practices and conventions. It is best to use completely fictitious data for this demonstration, although this is not essential. If you use real data, be sure to use pseudonyms. Be sure to include some data that produces difficult choices, and be very clear about the criteria that would be applied in these borderline cases. However, try not to include so many cases that meeting participants get confused by details. The example in the meeting materials section provides a guide. Allow plenty of time for questions.

15.6   Meeting Materials

1. Copies of institutional or departmental guidelines or regulations covering grades.

2. Copies of handout entitled "Administrative Issues for the End of Semester."

3. Copies of a calendar displaying all of the important jobs that the instructors are expected to do around the exam period, and the timetable that they should work to.

4. Copies of a flow chart that diagrams the official grade reporting process for your institution.

5. (Optional) Handouts or overhead projector slides for demonstrating final grade calculations.

Administrative Issues for the End of Semester [1]

1. **Scheduling Review Sessions**

   If you plan to hold review sessions for your class please request a room from the secretarial staff. Room requests can take up to a week to fill, especially during the busy examination period, so plan ahead.

2. **Before the Final Exam**

   Remember to schedule class time for students to fill out the student evaluations.

   Make sure that you have communicated the following information to every student in your class :

   - The date and time of the final exam.
   - The location of the final exam.
   - Details of any essential items that students must bring to the exam.
   - Details of any special items that students may bring to the exam.
   - The scheme that you will be using to calculate their final grades.

   Find out if there are any students in your section who will require an alternate exam. The main reasons for taking an alternate exam are:

   - The student must take another exam at the same time.
   - The student has four or more exams scheduled for the same day.

   Notify the course director if any of your students need to take the alternate exam.

   If you have made plans to leave town very soon after the final exam, please let the course director know well in advance. The reason for this is that if a student is unable to contact his or her instructor, he or she will often try to contact the course director.

   If you have arranged for someone to substitute for you in the grading session, please let the course director know well in advance, so that the grading effort can be coordinated efficiently.

   If you need to leave the grading session, for example if you have to take or give an exam in another course, please let the course coordinator know well in advance.

3. **Final Exam Day**

   You are required to proctor the final exam. Three reasons for this requirement are:

   - You need to physically identify all of the people from your section(s) who are taking the exam.
   - You will probably be the only instructor who is aware of any special needs that people in your class may have (e.g., students with documented learning disabilities). Your presence will ensure that these people are treated fairly.
   - You may have to deal with students who miss the exam, or who oversleep and arrive very late. You will probably be the only instructor in the room who will know whether or not people are missing from your section.

   When the people have started the exam, take attendance. Make a written record of the people from your section who are present, who have excused absences, and those people who are missing.

   When students begin handing in their exams, only accept exams from people in your own section(s).

   As students hand their papers to you at the end of the exam, check their names off. If you recorded a name, but did not check it off, then double check the papers that you were handed. Check with other instructors; they might have collected the paper by mistake.

   Before you leave the exam room, check the room (including under desks and inside waste paper baskets) for discarded exam papers.

---

[1]Reproduced from a handout used in the precalculus course at the University of Michigan in April 1998

4. **Cheating**

Dealing with cheating is difficult, personally demanding, and unpleasant. Forms of academic dishonesty that are of particular concern include

- looking at other people's work, and
- showing work to others.

If you observe what you consider to be cheating, tell one of the other instructors in the room. If the other instructor concurs, then get all of the instructors who are in the room to observe the person. If every instructor in the room agrees that there is cheating going on, then this is probably the case.

It will be essential to prepare a written case; this may include, for example, photocopies of questionable exams, written testimonies from other instructors and your own account. The mathematics department offices will be able to help you with the mechanics of reporting cases of academic misconduct.

Two final points :

- Many relatively innocent behaviors can look like cheating. Be absolutely sure that you have correctly identified academic misconduct before taking action.
- It can be very risky for you to apply "home-made justice" in cases of cheating. The official channels are cumbersome, but they are there to protect both you and the students; they should be used.

5. **Submitting Grades**

(a) **Before Submitting**

Before recording and submitting grades, it is important to decide in your mind why you are assigning these grades. In the vast majority of cases, the grade that you assign will be clear-cut, and your reasons for assigning that grade will be likewise clear-cut. In borderline cases, it is important for you to think carefully about the reasons for assigning one grade over the other, and once you have actually assigned the "borderline grade" to be sure in your mind why you have done so. Some reasons for this include the following.

- You may find that the decision of what grade to assign becomes much simpler once you have set forth reasons in your mind. Most people agonize over the grades they assign, and carefully thinking things through may help you to avoid some of this stress.
- If students come to complain or to argue over the grade that you have assigned, then you will already have thought things through. This may help you to avoid uncomfortable situations.

(b) **Submitting Grades**

Official grade reports are submitted to the Registrar's office on official sheets. The particulars of where and when to turn in your grade report sheets will be detailed in a memo from the department.

For the sake of your records, make a photocopy of your grade report sheet before turning it in, and keep the copy in a safe place.

If you plan to leave town before you are able to turn in your grade report sheet :

- Make sure that you have signed the sheet.
- Give the sheet to a trusted colleague to turn in on your behalf.

6. **Pass-Fail, Incompletes and Excused Absences**

The grades normally assigned are A, B, C, D, (each with possible + or -) and E. There are several other possibilities as follows :

W Official withdrawal

VI Visitor

These grades may appear, preprinted, on your official grade report sheet. Do not alter them.

(a) **Non-standard Grades**

The grades of I (incomplete), X (excused absence from both the final exam and the alternate), and NR (no report) should be given very rarely. These grades always remain on a student's transcript to indicate an irregularity. Unless you report a different grade within the appropriate deadline, these grades will lapse to an E.

You cannot give the grade of I to any student who has completed all of the work, nor can you allow a student to do extra work to raise the grade. An incomplete should be used only in the case of a student who is missing one portion of their grade (often a term paper or exam) and has arranged with the instructor to make it up.

Students who miss the final and the alternate should be assigned the grade X only if they have contacted you and presented an excuse satisfactory to you that their absence was unavoidable. All Xs should be given in consultation with the course director.

NR should be used only when a student who has never attended your class (or attended only at the very beginning of the term) appears on your grade sheet. A student who has taken exams, and who has only recently "dropped out of sight" should be given a grade in consultation with the course director.

(b) **Pass–fail**

A student may have elected to take the class as "pass–fail." Compute and record their grade as usual; the Registrar's office will convert your grade to either a pass or a fail. You will not normally know which students are taking the course pass–fail, and if you do know, it should not affect your grading decisions.

7. **Communicating Grades**

(a) **Returning Papers**

- **Via mail :** Interested students may provide you with a stamped, self-addressed envelope.
- **In person :** Some instructors announce a time when students may drop by and pick up their examinations

When returning papers to students, some precautions should be taken.

- Do not simply leave the exams in a public place, such as the corridor outside your office, for students to pick up as they please. Exams may be stolen or altered.
- If students want to take their exams, or if they want their exams mailed to them, make sure that they understand that as soon as the exam leaves your custody, no further changes can be made to the score on the exam.

(b) **Communicating Grades**

If students want to know their grades before official reports are mailed, they could give you a self-addressed stamped envelope or you could hold office hours at which time they could stop to check their grades. Another option is for you to E-mail the grade. If you like, you could invite students to write their E-mail address on the front of their exam if they would like you to E-mail their grade to them. Observe the following.

- The departmental offices do not give out grades to students.
- Do not post grades. This is prohibited by federal privacy regulations.
- Do not release grade information to anyone except the student concerned. Note that parents/guardians are not entitled to grade information.

8. **Handling Complaints**

The overwhelming majority of people in your class will never give you any kind of trouble over the grades that you assign. However, there is an increasing tendency for students to complain (sometimes through official channels) about grades. Some forms of preventive medicine include

- an excellent set of records;
- a rational, communicable grading system, which has been uniformly applied to all people in your class; and
- clear, well thought out reasons for the grade assignments that you have made.

If you can show that grades were arrived at by a rational system, uniformly applied, using accurate information, most complaints will disappear.

(a) Being Ready for Complaints

The best way to prepare for complaints is to have thought carefully about grade assignments before assigning the grades or speaking with students. In most cases, the reasons for the grade assignment will be clear cut and fairly mundane. Make sure that you are able to

- give specific reasons for the grade assignment (especially in "borderline cases"),
- demonstrate that objective criteria were used in determining the grade,
- demonstrate that these criteria were applied to all students uniformly,
- explain your decision within two minutes, and
- resist the temptation to digress into your own philosophies of education or testing.

(b) Common Student Arguments for Special Consideration

- "I require a certain grade to achieve a certain result."
  This kind of "argument" is usually based on rumors that the student has heard. An appropriate reply is to point out that every grade in the course was assigned in the same way, and that the student should discuss the situation with his or her academic advisor.
- "Is it possible for me to do any extra work to make up my grade?"
  This request is quite common, and must be answered with an immediate and firm NO. The reason is that every person in the entire course is entitled to the same treatment.
- "Can I see what the curve was? I'm only one point away from a <desirable grade>?"
  Some students may want you to re-grade everything, in the hope of finding a few points that will "put them over the top." There is no reason to feel compelled to do this. An appropriate response may be something along the lines of, "I consider this to be the grade that you deserve. I do not consider your grade to be border-line."
- "I really tried hard all semester. Don't I get some credit for that?"
- "I improved all semester. Doesn't that count for anything?"

In all cases, if you are able to demonstrate that objective criteria were used to determine the grade, and that these criteria were uniformly applied to everyone in the class, then most complaints will disappear.

(c) Belligerent Students

You may have an interview with a student who seems particularly combative, or who simply will not accept your answers. If you tire of the student's belligerence, it is suggested that you

- end the meeting, and
- arrange for either you and the student, or just the student, to meet with the course director.

After meeting with the student, and before contacting the course director, it may pay to write down, to the best of your recollection, exactly what was said during the interview.

(d) Improper conduct

It is exceedingly rare for students to attempt to bribe instructors. If you feel that a student is attempting to initiate a bribe (typically through some kind of joke or "Just for the sake of argument, say that ..."), then you should do the following.

- Terminate the interview immediately, and get the student out of your office.
- Find someone that you trust, and tell that person exactly what happened.

- Write down, to the best of your recollection, exactly what happened.
- Contact a course director as soon as you can.

9. **Changing Grades**

Grade changes are made by filling out and submitting a change of grade form, available from the secretaries. The only official reason for changing grades is to correct clerical errors, or to replace a grade of I or X. This is a convenient and unequivocal reply to students who simply request you to change their grade. Probably your best bet is to have thought carefully about grade assignments before assigning the grades or speaking with students. In any case, excepting correction of clerical errors, please talk the situation through with a course director before making the change.

10. **Post-Semester Responsibilities**

If students do not pick up their exams, the exams should be retained for one year after the date of the final.

Important Exam Period Dates [2]

April 1998

Monday	Tuesday	Wednesday	Thursday	Friday	Saturday
		1	2	3	4
6	7	8 **Course Meeting:** "Final Exam."	9 Exam rooms announced.	10	11
13 If you'd like a room for a **review session**, let the secretaries know this week.	14 Please announce **exam room**, **time** and **date** to your class each day.	15 **Course Meeting:** "Administrative Issues for the End of Semester."	16	17 In-class review. If you are behind, make an effort to finish today.	18
20 In-class review. **Last day** to take the **gateway** test.	21 **Last day of class.** **Math Lab closes.**	22 Official study day.	23 Official study day.	24 Final Exam. **8am–10am.** Grading until **finished**.	25 If you like, you may have some time for students to look at their exams.
27 **Final grades due** at the end of today.	28	29	30		

[2]Reproduced from a handout used in the precalculus course at the University of Michigan in April 1998

Calculating Final Grades [3]

The official breakdown for the final grade for this course is:

Exam 1 – 15%
Exam 2 – 20%
Final Exam – 25%
Homework – 25%
In Class Work and Quizzes – 15%.

The grade ranges for each of the uniform examinations will be determined on a course-wide basis. The grade ranges will be sent to you the day after the exam.

1. **Determining Your Uniform Grades**

   For each of the people in your class, calculate a *weighted uniform exam score*. This weighted score (W) is calculated from the students' scores on the mid-terms (S_1 and S_2), and the score on the final exam (S_F) as follows,

   $$W = 0.15S_1 + 0.2S_2 + 0.25S_F. \tag{15.1}$$

   As each exam was out of 100 points, W will be a number between 0 and 60. The *weighted uniform exam scores* for your class are used to determine the uniform grades for your class, according to the ranges that will be sent to you the day after the exam.

2. **Determining Your Homework Grades**

   Official course policy was to assign 15 homework assignments, each worth a total of 20 points. This makes a total of 300 points possible for the homework. Sum the cumulative score on homework for each student in your class, and plot the results on a number line from 0 to 300. Select the ranges for A, B, etc. on your number line so that the number of A grades is approximately the same as the number of A grades that your class has earned on the uniform grades, the number of B grades is approximately the same as the number of B grades that your class has earned on the uniform grades, etc.

   You will end up with something like:

Score Range	Grade
≥ 250	A or A–
≥ 235	B+, B or B–
≥ 210	C+, C or C–
≥ 180	D+, D or D–
< 180	E

3. **Determining Your Quiz Grades**

   Official course policy was to give 15 quizzes, each worth a total of 10 points. This makes a total of 150 points possible for the quizzes. Sum the cumulative score on quizzes for each student in your class, and plot the results on a number line from 0 to 150. Select the ranges for A, B, etc. on your number line so that the number of A grades is approximately the same as the number of A grades that your class has earned on the uniform grades, the number of B grades is approximately the same as the number of B grades that your class has earned on the uniform grades, etc.

   You will end up with something like:

---

[3]Reproduced from a handout used in the precalculus course at the University of Michigan in April 1998

Score Range	Grade
≥ 130	A or A–
≥ 120	B+, B or B–
≥ 100	C+, C or C–
≥ 90	D+, D or D–
< 90	E

4. Determining the Final Grades

For each student, determine their *weighted total score*, S_T. This is calculated from W (see (15.1)), the cumulative total on quizzes (Q), and the cumulative total on homework (H) as follows,

$$S_T = W + 15 * \left(\frac{Q}{150} \right) + 25 * \left(\frac{H}{300} \right). \tag{15.2}$$

The *final grade ranges* are then determined as follows:

(a) The lower bound for an A grade is calculated by adding the lower bound for an A from the uniform distribution (out of 60), 0.1 ($= \frac{15}{150}$) times the lower bound for an A from the *quiz* distribution, and 0.0833 ($= \frac{25}{300}$) times the lower bound for an A from the *homework* distribution.

(b) The lower bound for a B grade is calculated by adding the lower bound for a B from the uniform distribution (out of 60), 0.1 times the lower bound for a B from the *quiz* distribution, and 0.0833 times the lower bound for a B from the *homework* distribution.

(c) etc.

The uniform grades for the course may then be determined by comparing the result of (15.2) to the *final grade ranges* calculated above.

15.7 Suggested Reading

- Sharon Baiocco and Jamie DeWaters. *"Successful College Teaching. Problem-Solving Strategies of Distinguished Professors."* [6]

  Chapter 10 contains some advice for dealing with students who complain about grades.

- Steven Krantz. *"How to Teach Mathematics."* [69]

  Chapter 2 provides some suggestions on exams and grading (Sections 10 and 11), and on review sessions (Section 16).

  Chapter 4 is concerned with difficult situations, and it includes suggestions on cheating (Section 3) and "Begging and Pleading" (Section 11).

- Wilbert McKeachie. *"Teaching Tips."* [81]

  Chapter 6 describes some considerations for returning test papers to students and some measures that can help to moderate student aggression. Unfortunately, as McKeachie points out, student objectives in taking tests and instructor objectives in giving tests are not always well-aligned, so there are no easy answers here.

  Chapter 7 describes some ways in which students attempt to cheat. There is little here about preventing and handling cheating beyond the obvious (e.g., have students sit far enough apart so that they cannot easily see each others' tests).

  Chapter 8 describes some strategies for assigning grades. This is somewhat of a philosophical discussion, not a "How To" guide for assigning grades.

Chapter 16

Adapting Materials and Designing Your Own Meetings

16.1 Introduction

In developing this book, we have tried to identify a collection of experiences for training college mathematics instructors that is comprehensive and well integrated. This is appropriate, as teaching is a highly complex activity, especially in the setting of a student-centered interactive classroom, and it is somewhat artificial to treat teaching issues as isolated from each other.

The range of experiences that we have described in this book is consistent with our experience of the needs of graduate students and new faculty instructors at the University of Michigan. Naturally, other programs with different goals and methods may find that there are teaching issues that we have not substantially addressed, but which are nonetheless highly relevant. In this chapter, we hope to provide a framework for adapting materials to the specific needs of other programs.

16.2 A Process for Designing Your Own Meetings

We believe that instructor training meetings should be as lively as possible, and should actively involve the instructors as much as is possible. In a sense, the training meetings can, and perhaps should, serve as a model for how the instructors' classrooms should function. With this in mind, the "Lesson Planning Worksheet" included in Chapter 12 serves as a useful framework for planning a training meeting.

The major steps described in Chapter 12 are recounted below.

1. Read the section to be covered.

2. Identify the goals for the session.

3. Decide on a strategy to address each goal.

4. Begin time management.

5. Plan the student activities.

6. Write scripts for the mini-lectures.

7. Make an assessment plan.

8. Put it all together.

We feel that this sequence of steps provides an excellent framework for developing not only interactive lessons for undergraduates in mathematics, but also for planning interactive training sessions for instructors. It is our experience that as instructors become more experienced and proficient, they are able to abridge some of the steps in this process and still generate very effective lessons. We anticipate a similar effect when this process is used to generate training sessions. Given that trainers already possess a great deal of experience, this will probably happen very quickly.

No matter what process is used to generate training sessions, we have noted from our own efforts that successful training sessions usually share the following characteristics.

- There is a real need among the participants to have the training session.

- The participants have advance warning of the meeting, and understand what the meeting will be about before they attend.

- The session has a relatively small number of well-defined goals.

- The goals for the session describe what the participants will get out of the session, as opposed to merely delineating what issues will be "covered" during the session.

- There are opportunities for the participants to actively participate in, and contribute to, the meeting.

- The meeting leaders actually have something concrete and meaningful to say about the issues raised during the meeting.

- The structure of the meeting retains enough flexibility so that it is possible to address issues that participants raise within the meeting.

- The meeting ends on time!

The three points that seem to present the most difficulty are (1) informing participants ahead of time, (2) having something meaningful and concrete to say, and (3) ending the meeting on time. Points (1) and (3) are highly dependent not only on the organizational skill and dedication of the meeting leader, but also on the events that unfold during the course of the meeting. Appendix A describes some of the difficulties that can arise during a meeting, and offers suggestions for what leaders can do about them.

The remaining point is a stickier one—having something meaningful to say. As academics, and particularly as mathematicians, we have a natural tendency to want to speculate, generalize and think about situations in the abstract. Unfortunately, this is often not very relevant to the concerns of the participants—instructors who have given up either their personal time, or time that they could devote to other profession responsibilities. Some instructors may do this out of a sense of duty, but others do so in the hope that they can better their classroom situation with what they learn from your training sessions. With this understood, you have a responsibility to the participants: a responsibility to be able to respond to their concerns in ways that are relevant to them. This sounds like a very tall order (it is!), but for the training sessions to be valuable, it is necessary.

One of the difficulties here is that "trainers" are often picked because they have the reputation of being "good" teachers. One of the difficulties that both of us experienced in our initial training assignments was trying to help people make the best out of classroom situations that were well beyond our personal experiences. As "good" instructors, we are both adept at recognizing the potential for trouble, and taking steps to sort the trouble out before it becomes a problem. As professional educators we prepare thoroughly for class, and are rarely caught by an example that doesn't work properly, or an application of mathematics that we know nothing about. Visiting classrooms for the first time, we were occasionally amazed by how little the class we visited resembled our own ideals for an interactive mathematics classroom. This was not simply a matter of taste; when some *students* are yelling "SHUT UP!!" at the top of their lungs, while others loudly demand, "WHAT THE **** IS HE DOING NOW?" there is something seriously wrong! Having never experienced classroom environments such as these ourselves, we were initially at a loss to offer concrete advice to instructors. Experience has helped, but determined effort to seek out and utilize the resources that exist for helping mathematics teachers has also been a very fruitful activity. In particular, other volumes in the MAA Notes Series have been very useful in helping us to develop concrete solutions and advice for instructors who face profound difficulties (see suggested readings at the end of this chapter and the bibliography). These references can also serve as valuable sources of material that trainers can adapt to the specifics of their own program.

To illustrate the process of using resources to develop additional workshops, we outline the goals, assumptions about participants, agendas and materials for an extra meeting. The objective here is not to supply the same level of detail for running a meeting as was developed for the other chapters in this book, but to provide enough information to illustrate the kinds of considerations and methods that we use when planning new training sessions. We hope that this along with the steps outlined previously and the suggestions offered in Appendix A will serve as a useful framework for those who need to or wish to develop their own training sessions.

16.3   Example: A Workshop on Writing in Mathematics

Resources used to Develop the Workshop

The following is a list, with brief comments, of the references that we collected and used to formulate the goals and create the activities of the workshop. References are given in alphabetical order by author's surname.

- Patricia Shure, Beverly Black and Doug Shaw. "Michigan Calculus Instructor Training Materials." [103]

  This reference contains an example of a student's work (with no comments or explanations) that appears to indicate that the students has a clear grasp of the material. This reference also includes the same piece of student work with the student's attempts to explain his reasoning process. The version with the comments appears to reveal some serious problems with the student's understanding of the problem. This reference also includes a handout, written by Dale Winter, suggesting advantages that both instructors and students can realize from writing assignments.

- Annalisa Crannell. "How To Grade 300 Mathematical Essays and Live to Tell the Tale." [34]

  This article describes a system for grading writing assignments according to an eleven point checklist. Evidence (in the form of student grades on papers) is presented to support Crannell's claims that the grading system is "meaningful," "equitable to all students," "helpful to students' writing," and "time-efficient."

- George Gopen and David Smith. "What's An Assignment Like You Doing in a Course Like This?" [56].

  This article sets out a number of reasons for using writing assignments in mathematics classes, and describes some of the ways that an instructor can infer whether the students have a clear understanding or not. The article describes a theory for evaluating writing (Reader Expectation Theory), and includes a number of suggestions for instructors who wish to give feedback that will help students to become better writers.

- John Meier and Thomas Rishel. "Writing in the Learning and Teaching of Mathematics. MAA Notes # 48." [82]

  The first chapters of this book contain a wealth of ideas and advice on how instructors can use or include writing assignments in a mathematics course. This reference also includes examples from mathematics classes to indicate the characteristics of suitable writing assignments for students with different mathematical backgrounds. The later sections of the book deal with advanced topics, such as collaborating and learning from faculty in writing programs and designing major writing projects. These advanced topics could form the basis of a workshop for more experienced instructors.

- Andrew Sterrett. "Using Writing to Teach Mathematics. MAA Notes # 16." [109]

  This is a collection of articles describing individual teachers' experiences of using writing assignments in mathematics classes. Some of the most useful articles for this particular workshop are the ones that describe (1) exactly how instructors implemented writing assignments in their classes, and (2) what changes in students' responses to questions they observed.

Goals for the Workshop

After consideration of a collection of background material we decided to try to design a two hour workshop to achieve the following goals.

- To help participants gain a sense of what may be considered "good" mathematical writing, and to develop some confidence in their abilities to recognize it.

- To help participants recognize how writing assignments can develop information about students' understanding of mathematical concepts.

- To help participants to see that it is possible for students to appropriately manipulate symbols and equations, but at the same time have no underlying conception of the mathematics that they are carrying out.

- To provide participants with ideas for how to include writing assignments in their courses in ways that will not require major overhauls of the course—such as overhauls of goals, content and teaching methods.

- To help instructors understand the kinds of educational objectives that different writing assignments may be effective in achieving.

- To help instructors identify some of the issues involved in assessing students' writing.

- To equip instructors with practical ideas for helping students to become better mathematical writers through constructive feedback on writing assignments.

Assumptions about Participants

Now that we have a set of goals to achieve, we consider the instructors who may attend the training session. What do we know about what goes on in their classrooms, and how can this help us tailor the meeting to their interests? For example, if the instructors were all using writing assignments already, and convinced of their value, we may not explicitly address the first two goals, and instead concentrate on the variety of writing assignments that are possible, how they can be assessed, and how students can be helped through feedback. On the other hand, if the instructors have never even heard of writing in a mathematics class, then our emphasis would concentrate on why an instructor may include writing assignments and the variety that are possible. Assessment and feedback could form the content of a follow-up meeting to be held after the instructors had actually tried using writing in their own classes.

Based on our experiences, a reasonable set of assumptions about beginning graduate student instructors may include the following.

- Participants may not have ever considered that it is possible to exhibit "correct" mathematics with either no underlying conceptual knowledge, or with serious misconceptions about the mathematics.

- Participants may be worried about the amount of time that it takes to read and grade writing, especially if they have other commitments (such as a heavy teaching or course-work load, research or service on committees).

- Participants may be skeptical of their ability to motivate students to take the time and effort to write well.

- Participants may be concerned about the subjectivity of assessing writing—especially if the instructors are not very confident of their own abilities as writers.

- Participants may have no experience whatsoever—either as instructors or students themselves—of what to write *about* when writing in a mathematics course.

The Plan for the Workshop

As mentioned, the point of this section is not to fully describe how to run a two-hour workshop on writing, but to illustrate the process that we use to devise training sessions for mathematics instructors. In this section, we list each of the activities that we have planned for the workshop, which of the goals it addresses, and what assumptions about participants are involved, and an appropriate time frame for each activity.

- **What is good mathematical writing?**
  - **Procedures:** As participants arrive, hand them a sample of writing and ask them to comment on it (as though they were giving feedback to the author). Then ask them to rewrite the passage to make the meaning clearer. Participants can be encouraged to discuss in pairs, and when everyone has arrived, the meeting leader can have each pair contribute their thoughts. During this discussion, a list of "Features of Good Writing" may be compiled.
  - **Goals Addressed:** (1) To help participants to gain a sense of what may be considered "good" mathematical writing, and to develop some confidence in their abilities to recognize it.
  - **Assumptions:** Participants may be concerned about the "subjectivity" of assessing writing— especially if the instructors are not very confident of their own abilities as writers.
  - **Time Allotted:** 10 to 15 minutes.

- **What kinds of information can be developed from student writing?**
  - **Procedures:** This has the format of a mini-lecture. Show samples of students' work that exhibit deep understandings of mathematics and communicate these eloquently through writing. If possible, show contrasts between students' work with writing and without writing. If possible, show examples of work that are on the same topic. A very effective device is to show a piece of student work with only the symbol manipulations present, which appear to be perfect, and then overlay the student's written comments, which exhibit serious misconceptions.
  - **Goals Addressed:** (1) To help participants recognize how writing assignments can develop information about students' understanding of mathematical concepts. (2) To help participants to see that it is possible for students to appropriately manipulate symbols and equations, but at the same time have no underlying conception of the mathematics that they are carrying out.
  - **Assumptions:** (1) Participants may not have ever considered that it is possible to exhibit "correct" mathematics with either no underlying conceptual knowledge, or with serious misconceptions about the mathematics.
  - **Time Allotted:** 10 minutes.

- **How do you include writing in a mathematics course?**
  - **Procedures:** This is a combination of mini-lecture and activities. Provide instructors with a short lecture that briefly described several different kinds of writing assignments (autobiographies, "EXPLAIN" questions, journals, long-term projects, etc.) and when it may be appropriate to use these in a course. Give participants an opportunity to work in pairs or small groups to examine the topics that they will be treating in their own classes during the next week or so, and to think of ways that they could include writing assignments. The meeting leader can emphasize, as the groups work, the importance of making the writing assignments something that the students will feel is significantly contributing to their learning of mathematics. At the end, the leader can facilitate a discussion where each group reports its ideas.
  - **Goals Addressed:** (1) To provide participants with ideas for how to include writing assignments in their courses in ways that will not require major overhauls of the course. (2) To help instructors understand the kinds of educational objectives that different writing assignments may be effective in achieving.
  - **Assumptions:** Participants may be skeptical of their ability to motivate students to take the time and effort to write well.
  - **Time Allotted:** About 30 minutes total.

- **Break**—refreshments and informal discussion of the possible roles of writing in addressing difficulties with mathematics classrooms.

- **What are writing assignments good for?**

  - **Procedures:** This is a combination of mini-lecture and discussion. The idea is that this is strongly linked to what the participants were doing immediately before the break, and is fast-paced enough so that the meeting gets going again. A handout or overhead listing pros and cons of various kinds of writing assignments can be very helpful here.

  - **Goals Addressed:** (1) To help instructors understand the kinds of educational objectives that different writing assignments may be effective in achieving.

  - **Assumptions:** None.

  - **Time Allotted:** About 5 to 10 minutes.

- **Assessing Writing Assignments**

  - **Procedures:** Introduce instructors to some of the possible options for assessing writing, e.g., Annalisa Crannell's guidelines. Ensure that instructors appreciate that grading writing may seem more subjective or more nebulous than grading "algebraic manipulations," but that they can cope by developing check lists of criteria that they think important and using these to guide their assessment. Point out that consistency is sometimes a problem, especially since students will have different levels of ability in writing, as well as different levels of ability in mathematics. Make sure that instructors understand that a check list is a good way to help mediate some of the problems with inconsistency. If time permits, suggest strategies that help students to improve their writing abilities. For example, rewrites and re-submission policies. Point out some of the problems with assessing writing assignments. For example, if you have a rewrite policy, some students may think that just by responding to the specific comments on the writing assignment, but not substantially revising their work, they will get full points.

  - **Goals Addressed:** (1) To help instructors identify some of the issues involved in assessing students' writing.

  - **Assumptions:** (1) Participants may be worried about the amount of time that it takes to read and grade writing, especially if they have a heavy course-work or thesis research commitment. (2) Participants may be concerned about the "subjectivity" of assessing writing–especially if the instructors are not very confident of their own abilities as writers.

  - **Time Allotted:** About 20 minutes.

- **Helping students to become better writers—Constructive Feedback**

  - **Procedures:** Describe to the participants the ways that students tend to read instructor's comments. For example, some students read the comments as an ongoing story, rather than thinking about how each comment relates to the text it is attached to. Likewise, discuss how students tend to respond to comments, e.g., by making minor modifications of text, rather than substantial revisions. Discuss some of the ways that students use vagueness, dangling pronouns and the passive voice to "cover up" areas that they don't really understand, and ways that instructors can detect this. Present instructors with some examples in which students seem to be avoiding making a bold and direct statement of what they think the mathematics means, and ask instructors to discuss the possible comments that they could write. At some point, make sure that the motive for uncovering and responding to this kind of writing is clear—that is, to help students understand the mathematics more thoroughly.

  - **Goals Addressed:** (1) To equip instructors with practical ideas for helping students to become better mathematical writers through constructive feedback on writing assignments.

- **Assumptions:** (1) Participants may have no experience whatsoever—either as instructors or students themselves—of what to write *about* when writing in a mathematics course. (2) Participants may be concerned about the "subjectivity" of assessing writing—especially if the instructors are not very confident of their own abilities as writers.

- **Time Allotted:** About 20 minutes.

- **Wrap-up and assessment of workshop by participants**

16.4 Some Practical Tips for Planning Meetings

1. Make sure that you can actually do all of the critical parts of the meeting in the amount of time that will be available for the meeting. A paramount criticism of meetings is that they do not end on time. On the other hand, there will be critical information and issues that need to be discussed.

2. Try to include opportunities for meeting participants to do something active when this is possible and appropriate.

3. Try to start meetings with a warm-up or with some of the easier meeting items. This can help if some people are a little late to the meeting. Right after the beginning of the meeting, most of the participants will be paying the closest attention, and this is a good time to discuss important issues or have the meeting participants work on difficult activities.

4. Consider different approaches for addressing meeting objectives, and decide which you can use effectively. As mentioned, simply presenting information in the form of a lecture is not always to promote change in teaching methods. If you are familiar with active and cooperative learning strategies you may find that many of these techniques can be readily applied to activities in training meetings.

   A few alternatives to lecturing include the following:

   - Pose a question to participants, and have then discuss their answers in small groups. For example, "What does it mean to *motivate* students?" can be used to stimulate discussion.

   - After introducing an instructional technique, ask participants to plan a lesson using that technique. Ask participants to discuss the lesson plan with a neighbor.

   - Use role-plays to make or illustrate points.

   - Instead of presenting lists of considerations, try to turn the list into a worksheet. See the chapter on writing lesson plans for an example. Have participants work through the sheet instead of presenting the lists.

   - Invite guest speakers in the form of experienced instructors.

   - Instead of having a traditional training meeting, have participants research topics themselves, and give short presentations during the meeting.

16.5   Suggested Reading

- Bettye Anne Case. *"Responses to the Challenge: Keys to Improved Instruction. MAA Notes #11."* [23]

  This book is in two sections. The first section describes responses to a large-scale survey on instructional practices at major universities. This was first published in 1989, and the data may no longer be very relevant. The first section also includes statements from various academics on what skills and capabilities teaching assistants and part-time instructors should have, and what kind of training opportunities should be provided for them. Again, this information may be dated, but it may be a helpful starting point. The second half of the book is a collection of the "Instructor Guides" that were used, circa 1989, in a number of major mathematics departments that employed large numbers of graduate teaching assistants. Although some of these guides are deeply influenced by the fact that lecturing was the dominant method of mathematics instruction in those days (so much of the discussion centers around "effective lecturing"), these guides serve as an excellent indication of the kinds of information that beginning instructors will not usually know.

- Bettye Anne Case. *"You're the Professor. What Next? Ideas and Resources for Preparing College Teachers. MAA Notes #35"* [24]

  This book is a collection of statements about what is needed to teach mathematics, perhaps with an emphasis on elementary mathematics, at the college level. Statements are included from very prominent mathematicians. There are also a number of descriptions of teaching training programs and doctoral programs from universities with large graduate programs. All in all, the focus of this collection of articles may be summarized as a variety of perspectives on what kinds of teaching capabilities are needed by faculty, and what Ph.D. programs ought to be doing to equip their students with these capabilities.

- Leo Lambert and Stacey Lane Tice, eds. *"Preparing Graduate Students to Teach."* [70]

  This book is a collection of relatively brief descriptions of several graduate teaching assistant training programs at prominent universities. Most of the descriptions are rather telegraphic in nature, due to space constraints, but give reasonable outlines of the topics included and scope of the training programs. Not all of the training programs described in this book are in mathematics.

Chapter 17

Classroom Visits

17.1 Class Visits and Professional Development

17.1.1 Overview of Class Visits

An important component of a professional development program is a system of visits to new instructors' classes. These visits take a substantial amount of time and effort, but they afford many benefits that cannot be gained from other aspects of a professional development program. Through class visits, instructors can get concrete, focused, and personal feedback. Through the sharing of sensible and helpful advice, the course coordinators can gain the respect and trust of their instructors. Visits keep the coordinators connected with the classrooms: the nature of the students, the classroom layout, and developing problems course—wide or with individual instructors. Finally, visits are a valuable tool for assessing and improving the professional development program.

The bulk of the chapter is an explanation of three types of class visits. The types of visits are observational visits, student-feedback visits, and peer visits. The observational visits are conducted by a course coordinator, and are detailed in Section 17.2. The student-feedback visits are also conducted by a course coordinator, but a similar less effective method can be conducted by an instructor in his or her own class. These types of visits are detailed in Section 17.3 and Section 17.4 respectively. Peer visits allow instructors to give and receive feedback, and don't involve the course coordinators. These visits are detailed in Section 17.5. These sections describe the visits, include some sample forms and artifacts from visits, and list some of the advantages and disadvantages of each type of visit. These visits can be implemented separately, or combined into a structured program of visits. Such a program is discussed in Section 17.1.3.

The chapter concludes with a section on potential problems a course coordinator may face when implementing a program of class visits. The section includes suggestions for avoiding some of these problems, and a list of sample situations to think about before implementing class visits.

17.1.2 Possible Goals of Class Visits

- Goals for instructors:

  - To help instructors develop and articulate teaching goals.
  - To point out positive aspects of instructors' teaching, in order to give them some things to "hang their hats on."
  - To give instructors an opportunity to voice and discuss their concerns privately.
  - To identify areas of classroom practice on which the instructors need to concentrate their efforts towards improving.
  - To expand instructors' range of teaching strategies.
  - To give inexperienced instructors opportunities to interact with, observe, and receive feedback from more experienced instructors.

- To encourage instructors to reflect upon and assess their own teaching choices, strategies, and effectiveness.

- Goals for trainers:

  - To look for teaching ideas that can be shared with other instructors.
  - To observe each instructor's approach to implementing the teaching techniques covered in the professional development program.
  - To identify particularly strong instructors, who would be good candidates for peer visits.

17.1.3 Ideas for Using Visits Throughout the Semester

The types of visits explained here can be combined into a fully-developed program that would utilize the advantages of each type of visit. If the course coordinator chooses to develop such a program of visits, he or she needs to be sensitive to the amount of time and effort required to execute the program. The advantages of the different kinds of visits should be weighed against the disadvantages of the amount of time required, and forethought should be given to the interplay among the types of visits. If the team of facilitators and the instructors feel that the program of visits contains redundancies and unnecessary time commitments, then receptivity to feedback can be greatly diminished, and the value of the program can be lost.

One way to structure such a program of visits would be:

Two weeks into the semester: the course coordinators conduct observational visits.

Midway into the semester: the course coordinators or institutional instructional consultants conduct Student Mid-semester Feedback (alternatively, the instructors can collect their own student feedback).

Last third of the semester: the instructors are paired and asked to conduct peer visits.

If done correctly, this program of visits will give rich and varied feedback to the instructors and to the course coordinators. The program begins with an observation by the course coordinators very early in the semester to give immediate support to new instructors, and to diagnose potentially serious problems. The Student Mid-semester Feedback allows the instructor to obtain facilitated feedback from the students at a point in time when the students have had enough experience in the class to generate informed opinions, but also when there is enough time left in the semester to affect meaningful changes. Finally, the program concludes with peer visits. By the end of the semester, the new instructors should have gained enough understanding of teaching practice and feedback collection to begin to participate in the assessment of their own teaching and the teaching of others. Thus structured, the program is designed to move the instructors from being dependent upon a supervisory source of feedback to being independently able to collect their own feedback and reflect upon their own teaching.

The value of the visits is maximized when the visitors can observe actual teaching, and when the students are not overly distracted by unusual circumstances. Therefore, when scheduling the program of visits, the course coordinator must avoid using the class periods right before exams, immediately after exams, and in close proximity to lengthy vacations.

17.1.4 Special Considerations for Working with Novice Instructors

Observing and providing feedback to very new and inexperienced instructors can present special challenges. In compiling this chapter, we have assumed that a major goal of a program of classroom visits is the improvement of instruction, according to the standards, accepted practices and traditions of your institution. In particular, we have not advocated using a program of classroom visits as a vehicle for confronting instructors' customary teaching practices, or the beliefs about learning and teaching that underscore those practices. While it certainly is possible to confront instructors' beliefs during a program of classroom visits, we have adopted an approach of attempting to work with or around instructors' existing belief structures, partly because research [111, 72] (and personal experience) demonstrates that these beliefs are often very strongly held and highly resistant to change.

When conducting any observation and providing feedback, it is important to bear in mind that the instructors' beliefs about learning and teaching are involved, and that these beliefs may make the instructor more or less receptive to your feedback, depending on whether it is compatible with the instructor's existing ideas [122]. When providing feedback to novice instructors, we have found it useful to bear the following three guidelines in mind.

1. Even though novices do not have a great deal of direct experience as classroom instructors, they generally do have a lot of experience of classrooms, and will often have quite definite and strongly held ideas of what "works," what does not work, and what is permissible in a classroom [38, 72]. Often, this is based on the novice's memories (or current experiences if the instructor is a beginning graduate student) of what "worked" for them as a student [20], and may not be either accurate or appropriate for the students currently under instruction [105].

2. Even though some novices feel that they know a lot about classrooms, they may not yet feel comfortable in the role of classroom instructor [122]. In particular, as some authors [69, 98] point out, some classroom events may be very frightening for novice instructors, and the thought of having someone observe the class scarier still. Exacerbating this already tense situation is the fact that some instructors may equate a classroom visit with an evaluation of them as a person [107].

3. Finally, although some novice instructors show a remarkable insight into the personalities and processes that take place in their classroom, some novices have yet to develop a sophisticated and comprehensive framework for analyzing and making sense of classroom situations [122]. There is some evidence [122] to suggest that at least some instructors have a very dualistic [88] view of the nature of teaching, tending to see their actions as either "right" or "wrong" according to some external, fixed system of reference.

Some immediate corollaries of these observations are listed below.

- It is not safe to assume that you are dealing with a "blank slate" when providing feedback even to very inexperienced instructors. One way to uncover the beliefs that the instructor holds is to ask some probing questions when you first make contact to set up the classroom visit.

- Even if you believe that your observations and suggestions are penetrating, brilliant and exactly what this instructor needs to do, the instructor may reject them simply because he or she cannot reconcile the suggestions with his or her current beliefs about what "works" or is appropriate for a college mathematics classroom.

- We have both noted, as we have become more experienced at providing feedback to instructors, that we are less and less comfortable with simple, dualistic notions of "right" and "wrong" (at least in reference to teaching practices). However, even the most promising novice instructors that we work with usually *need* very definite and absolute advice because they have not yet developed the broad perspective on teaching and learning that can accommodate a wide range of ideas. Instead, they are principally concerned with what goes on within their own classroom. For this reason, we have continued to advocate that novice instructors be provided with concrete, straight forward suggestions over complicated, relativistic analyses. For this reason we use simple section headings in our reports (e.g., "What is going well").

- Nothing dampens a person's enthusiasm for a new job more than to be told by a so-called expert that everything is being done terribly. When visiting the classrooms of novice instructors we make a careful and deliberate effort to "catch them doing things right" [11]. As an added bonus, this recognition that the novice has got a lot of things "right" can make them more receptive to your suggestions [22].

17.1.5 The Impact of Class Visits on the Professional Development Program

A well-functioning program of class visits will not only benefit the individual instructors getting feedback, but it will benefit the course coordinators as well. If the course coordinators are regularly visiting classes, they stay in touch with the nature of the student body, the physical layouts of the classrooms, and potential problems with individual instructors or the course overall. In addition, the coordinators gain a renewed appreciation for the day-to-day issues that new instructors face.

The feedback gathered by a program of class visits can also be used to customize, improve and fine-tune the overall professional development program. The feedback can alert the course coordinators to issues requiring immediate attention during course meetings. Feedback can be used to customize some of the more open-ended meetings in this volume, such as the meeting on "Dealing with Difficult Instructor—Student Situations" in Chapter 14 and the meeting on "Establishing and Maintaining Control in Your Classroom" in Chapter 9. The feedback can also be used to assess the overall effectiveness of the training program, and bring ideas for changing future versions of the professional development program.

Finally, feedback from class visits can be distilled and disseminated among all the instructors. For example, the following handout was created from the reports of Student Feedback visits conducted during one semester at The University of Michigan. The idea behind creating this document and distributing it to all instructors in the course was to indicate some of the positive instructor qualities on which many classes had commented. It was hoped that this may give the instructors some idea of what was making successful classes succeed. A similar list of suggested changes could have been compiled and distributed to instructors as well. Such a list might have been even more effective at getting instructors' attention.

The method used to generate this handout was to scan through the reports written for all classes visited, and try to identify themes. Under each of those themes, comments were added from specific reports, and identifying information was removed from the comments.

Encouraging Participation

During early semester feedback of 17 precalculus classes, the following instructor characteristics were identified by the students as characteristics which made class a better learning experience. There were many characteristics given, the ones listed below (along with comments from students) represent characteristics that were identified in several different classes, each taught by a different instructor.

- **Enthusiasm**

  - Always enthusiastic, eager and willing to answer questions.
  - Enthusiastic. He is not boring. He tries to use everyday examples to make the problems more interesting.
  - Enthusiastic about course and it helps to keep us interested.
  - He has good initiative and wants us to learn. Enthusiastic.

- **Being knowledgeable and able to explain things in more than one way**

  - Patient and breaks everything down simply (sometimes).
  - Instructor is knowledgeable and wants us to understand. Asks class after explaining problem, "Does everybody understand?"
  - The instructor's ability to explain things like questions on homework.

- **A friendly, patient, non-threatening manner**

  - <NAME> is very friendly outside of class.
  - Very patient when we don't understand.
  - Very friendly and approachable.
  - Even tempered, patient. Doesn't get upset, is very calm.
  - He cares about the students.
  - The teacher is easy to talk to.
  - She is approachable—can ask questions about the work.
  - Cool guy, very friendly, especially during office hours. Open to questions.
  - He creates a friendly environment that promotes learning.
  - Always jokes around, lightening moods and math frustration.

- **Making an effort to get to know people in the class**

  - He treats us like people rather than students—knows our names.
  - Tells us about himself (personal history).
  - The instructor seems to know our names, and uses them in class.

- **Letting the people in the class know that you are human**

  - He is prepared to admit that his knowledge is not boundless (e.g., economics).

- **Being supportive of the people in your class**

  - She is very encouraging—she compliments us on the quality of our homework.
  - Even when we're stuck, she tells us that she knows we can figure it out.

- **Being sensitive to the people in your class and their concerns**

- Caring personality shines through by adapting quiz scores, office hours and E-mails.
- She is very responsive to our concerns. For example, e-mailing gateway scores.
- He is accommodating—if you cannot make it to office hours you can come another time.
- He'll listen to us when we are swamped with other class work and exams for other classes.
- Considerate of workload. Extends deadlines when really needed.

- **Running a classroom where active participation is the norm**

  - Groups are good—because people can help one another learn the material—if you have a good group.
  - Gets everyone involved. Calls on people in class.
  - Calls on individuals to do problems on the board.
  - Class is interactive. Gets everyone involved.

- **No nasty surprises**

  - Concrete schedule and course activities well planned.
  - Every day he puts on the board what is coming up during the week.
  - Appreciate warning of quizzes.

Some Strategies to Encourage Participation

- **Finding (and using) alternatives to instructor-centered lecturing.**

    - Interactive discussions
    - Using questions to stimulate people to think about the material
    - Having people from the class explain the material
    - Having people in the class work together to learn the material

- **Finding (and using) activities in class where students are required to participate.**

    - Calling on people by name
        * To answer your questions
        * To explain their solution to a problem to the class
        * To present or explain a concept that they understand to the class
    - Having students work together, and present their results to the rest of the class

- **Finding out what the people in your class are interested in, and targeting your treatment of the material to those interests.**

    - Distribute and analyze student data sheets
    - Choose applications that are appropriate for your students' majors, such as business versus physical applications
    - Make activities that reflect campus events

17.2   Observational Visits

Many instructors will be familiar with this kind of class visit. It simply involves an observer attending the class, recording the significant events that take place, and then meeting with the instructor to give feedback. This is perhaps the least intrusive kind of class visit described here.

17.2.1   A Procedure for Conducting Observational Visits

1. **Setting Up.** Well in advance of the intended visit, contact the instructor to be visited, and set up a day and time for the visit. It helps to choose a day for the visit when the instructor will actually be teaching, so days when the instructor plans to give lengthy quizzes or exams should be avoided. Occasionally, instructors will forget about the visit, and then panic when they arrive to teach and find the observer there. A reminder the day before the class visit can help to avoid this situation. Contacting the instructor before you visit the class also gives you a chance to find out what the instructor has in mind for the lesson, and whether or not the instructor has any specific points that he or she would like feedback on. Some authors, e.g., Sorcinelli [107], emphasize that the information gathered when setting up the visit can make the observation much more meaningful for the observer, particularly if the class observed is very strongly connected to events or topics from other class meetings that the observer has not seen.

2. **The Visit.** Arrive early, and begin to note the physical layout of the room, as well as the behavior of the students and the instructor. Before the class begins, set up a time and place to meet with the instructor for up to an hour to review the observations. During the class, record as much data as possible. Some helpful rubrics are drawing a classroom map, jotting down a detailed script of what is said and done in the class, taking lecture notes from a student's point of view, or simply noting the significant events that take place. Avoid value judgments at this stage, and take only detailed objective data. In addition to noting what the instructor does, be sure to note what the students are doing. Do they seem engaged? Are they paying attention? Are they reading the newspaper and talking? The following are suggestions on what to look for as an observer, and they are adapted from the video "Observing Teaching" [26].

   - Note the physical layout of the classroom.
   - How does class begin?
   - Note the patterns of talk between instructor and students and among students.
   - Note the patterns of movement.
   - Note what is written on the blackboard, whether it is readable, and how long it is left on the board.
   - Note what types of questions the instructor uses. Note how much time elapses before the questions are answered.
   - Does the instructor give clear directions for activities?
   - Note the mannerisms of the instructor and students, including eye contact and tone of voice.
   - How does the class end? Is there a summary?

   Examples of the kinds of observations made are included in Section 17.2.2. At the end of the lesson, give the instructor an encouraging word (the simple phrase "Nice job!" has set many instructors' minds at rest), thank the instructor, and confirm the appointment to review the observations.

3. **Writing the Report and Planning for the Meeting.** Use the notes and maps made during the observations to compile a report. One possible format for this report is to list observations and conclusions drawn from observations under the following headings:

   - **What is going well in the class.**
   - **Things to watch out for.**
   - **Suggestions for action.**

Be concrete and specific. Vague comments simply confuse instructors, and can leave them feeling that the visit was a waste of time. At this stage it is appropriate to make value judgments based on the objective data. Examples of reports compiled using this format can be found in Section 17.2.2.

We note that if taken at face value, the headings that we have suggested might seem to imply that there is some universal standard for assessing the quality of or problems with teaching. (See Section 17.1.4 for further discussion.) We do not mean to imply that any such standard exists. In any given institutional and instructional setting, there are almost certainly a wide range of teaching behaviors that could be considered to be "effective," and a classroom that featured these teaching practices could be thought of as "going well." Many institutions have guidelines, suggested practices or perhaps even ready-made classroom observation forms that you can use both to gather information during the observation and to select the items that you plan to mention in your write-up. In Section 17.1.5, we included a document on ideas for encouraging participation as an example of one way in which a program of visits can contribute to an instructional program as a whole, rather than just to individual instructors one at a time. Such documents can also be a source of ideas for what to include or focus on when you compile your write-up. A wide range of references with similar guidelines and suggestions exist (e.g., [51, 124, 95, 43, 9, 38, 69, 98]) reflecting a variety of perspectives on teaching and learning.

4. **The Meeting.** Meet with the instructor at a mutually convenient time and place to discuss the class observed. Meet with the instructor while the class session is still fresh in his or her memory, and before the next class session if possible. To remove some of the intimidation inherent in the situation, it is best not to meet in your office. Instead, choose a public place that is not specifically populated by members of your department, such as a coffee shop but not the departmental lounge. The site should be private to the instructor's peers, yet public enough to be non-threatening.

Thank the instructor for allowing the classroom visit to take place, even if he or she had no choice in the matter. It can be helpful to acknowledge that a classroom visit, no matter how well-intended, can be an intimidating and stressful experience for the person being observed.

Begin the dialogue by asking the instructor, "What were your goals for that class session, and what do you think went well for you in that class?" This is a good opportunity to get the instructor to think about his or her teaching philosophy, and how to connect this to teaching practice. Other questions that can stimulate discussion include, "Was that class session typical?" "Can you share with me how you decided to organize that class?" and simply "How do you feel things are going in your class?"

It is usually the case that the observer will have to insist on talking about what is going well in the class. Instructors are usually eager to read the "criticisms," and sometimes ignore everything else, unless the observer insists on discussing the positive aspects of class first. However, if the positive aspects are emphasized at the beginning, the instructor may have less of a tendency to become defensive.

Be sure to corroborate the warnings and suggestions with the objective data. Emphasize two or three strengths on which the instructor can faithfully rely, and emphasize two or three main areas on which the instructor should concentrate efforts towards improvement. If the observer keeps in mind the following characteristics of constructive feedback found in [8], the feedback can be more effective.

Characteristics of constructive feedback:

- It is *descriptive* rather than evaluative.
- It is *specific* rather than general.
- It is *focused on behavior* rather than on the person.
- It takes into account *the needs of both the receiver and giver of feedback.*
- It is directed toward *behavior which the receiver can do something about.*
- It is *solicited* rather than imposed.
- It is *well timed.*
- It *involves sharing of information.*
- It involves the *amount of information the receiver can use* rather than the amount we would like to give.

- It concerns *what is said and done*, not how, not why.
- It is *checked to ensure clear communication*.
- It is *checked to determine degree of agreement from others*.
- It is followed by *attention to the consequences of the feedback*.
- It is an important step toward *authenticity*.

In addition, the listening and speaking techniques outlined in the handout entitled, "Techniques for Clear Communication" in Chapter 14 can be modified to the situation of an observer and an instructor in order to ensure more effective feedback.

17.2.2   Examples of Notes and Reports from Observational Visits.

- Instructor is present before class begins, handing out quizzes and gateways.

- Collects homework before class.

- Jokes with class. "We'll figure out a suitable punishment."

- Announces and writes up important info, e.g. exam.

- There is quite a bit of talking during quiz; some students actually checked their answers with others before turning it in.

- Nothing for people to do if they finished early (almost everyone did).

- Gives students a chance to ask about HW (starts at 1:21 pm, ends at 1:32pm)

- Instructor asks for suggestions from the class.

- It is hard to hear responses/questions from the front of the room

- Helps students with exam (explains assumption of squares).

- Instructor seems to be trying to learn people's names.

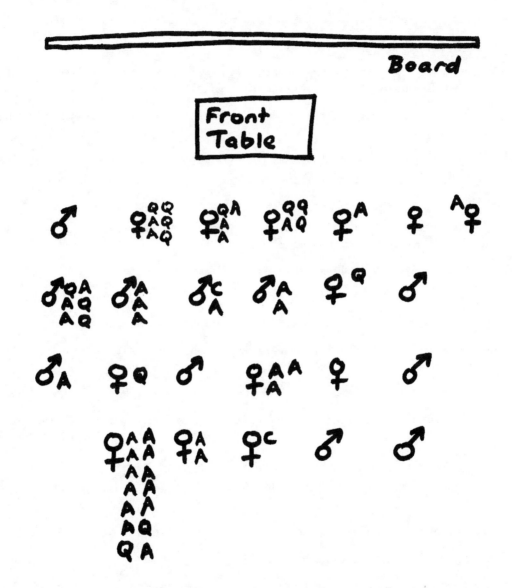

Key: Q = asked instructor a question

A = answered instructor's question

C = made comment.

Instructor :
Class :
Date :
Observer :

What is going well in class

- The instructor arrives before class and takes care of administrative issues before class begins.

- The instructor announces and writes out important information.

- The instructor deals with "Smart Alec" remarks.

- The instructor gives people in the class a chance to ask about homework problems, and does a good job of keeping the amount of class time devoted to this under control.

- When working through the homework problem, the instructor has people from the class supply suggestions.

- Throughout the class, the instructor makes the most of opportunities to model how they could explain their working on exams.

- The instructor seems to know people's names, and uses them in class.

- When explaining at the board, the instructor checks with the class at each stage of the problem to make sure that everyone understands.

- The instructor seems friendly and confident.

- The instructor identifies pitfalls and points them out to the class (e.g., can't simplify $\sqrt{8} + \sqrt{13}$).

- The instructor seems responsive to comments and questions from the class.

- The instructor gives very clear instructions on how and where groups should form, and what problem they should work on — he writes the problem and page number on the board.

- The instructor gives clear directions on what he expects the class to do during the group activities.

- The instructor sets a definite pace for the group work — gives the groups a definite time frame to work in.

- While the groups are working, the instructor visits each group to see how they are doing.

- When wrapping up the group exercise, the instructor has people from the class describe the answers that they got, rather than explaining the answers himself.

- When he has something important to say, the instructor is prepared to call for attention.

- The instructor notes important points on the board.

- During the second group exercise, the instructor tries to keep people working on the problem assigned (guy who wanted to learn how to program his calculator).

- The instructor has a problem for people to go on with if they finish early.

- The instructor has a very friendly manner. He is neither overbearing nor bossy, even when getting students back on track.

- The instructor's explanations are very concise.

Things to watch out for

- There was a lot of talking and general lack of discipline during the quiz. Some people actually compared their answers before handing their papers in.

- Nearly everyone finished the quiz early, but there was nothing for them to go on with.

- It was very hard to hear what people at the front of the room were saying. This was frustrating when they answered a question, but you couldn't hear what was said.

- It was difficult to see the bottom of the board.

- When the instructor asked questions that were directed to the class as a whole, there was a tendency for the same few people to answer each time.

- People in the groups at the back of the room would often give answers that were complete, correct and well explained, but the instructor would miss them (especially Jennifer and the woman in the group right at the back).

- Some of the groups were not sitting close enough to work together easily.

Suggestions for action

- Look for groups that are obviously not working on the assigned activity, and encourage them to cooperate. Some ideas for things that you could say to them :

  - You could ask them which parts of the activity they have managed to work out, and have them explain one of their answers to you.

  - You could tell them that you will be calling on their group to supply an answer when you wrap up the group activity.

  - You could simply tell them to get to work on the problem at hand (without being too heavy handed about it).

- If the group activity is a long one, with multiple parts, you can break up the group work by asking if all groups have finished, say, parts (a) and (b). If this is so, then stop the class and discuss parts (a) and (b). When finished, have them go back to the rest of the problem. This may help slow groups keep up, and give you another way to set some kind of pace in the group activities.

- When someone at the front of the class asks a question, check to see that everyone in the class heard the question. If not, repeat the question for their benefit.

- Enforce a higher level of discipline on quizzes. Have an exercise that they can go on with if they finish early.

- When you are asking questions, move around the room. This may give you a chance to hear contributions from more people.

- If you see someone sitting out from their group, encourage them to move their seat.

17.2.3   Advantages and Disadvantages of Observational Visits

Advantages of Observational Visits

- This kind of visit is possible very early in the semester, and can provide new instructors with advice and observations on their teaching and suggestions at a time when these can have the greatest effect on the conduct of the class.

- If the report contains concrete suggestions for action, then the instructor will have something concrete on which to work.

- The realization that he or she is doing something right usually has a very positive effect on the instructor's morale, confidence and interest in the course.

- This kind of visit is relatively unobtrusive. No class time is devoted to the visit or visitor; ideally the class observed will be "business as usual." This can be an important consideration when the syllabus is very full.

- If the observer is an individual that the instructor respects, and is not evaluating the instructor in some way, then the instructor may be quite interested to hear what the observer thinks.

Disadvantages of Observational Visits

- The report will be most effective if it is accurate, specific and concrete. It *is* difficult and demanding to observe, record and report with the level of detail required.

- Although a lot of attention to detail and specificity is required, the report needs to be written with economy, so that the instructor is not overwhelmed or dismayed. Writing reports that are fair, informative and not overly long requires some skill.

- Because this type of class visit requires a lot from the observer, many observers may be needed in order to visit all of the instructors without becoming overwhelmed by the amount of time and energy required.

- Some instructors may consider this to be an *evaluation* of their teaching, and react strongly to any perceived criticism.

- Some instructors may not take the observer's efforts and the results of your observations seriously.

- Some instructors may have the attitude that whatever happens in the classroom is between them and the students in their class; outsiders and visitors are not welcome.

17.3   Student Feedback Visits

The process called "Mid-semester Feedback" is a method that uses group discussion amongst students to provide feedback to their instructor. Over the last three years, the method has been widely used in math classes, with very positive responses from both the instructors and the students concerned.

A reasonably substantial literature (see the suggested readings at the end of this chapter) exists to describe the benefits of this type of classroom visit, although we note that it is probably a less frequently used form of classroom assessment than the direct observation visits described earlier. While this form of class visit appears to be more elaborate and difficult to conduct than a direct observation, we found that with minimal training, we were quickly able to conduct this kind of visit in a highly efficient and productive manner. Any instructor who is familiar with the use and facilitation of in class cooperative group work already has most of the skills required to expertly conduct a student feedback visit. In addition, we found that writing reports based on student feedback was a much quicker and easier process than compiling reports from our own observations, and that much of the information developed would have been unattainable through observations made by an instructional consultant.

Students have expressed greater satisfaction with mid-semester feedback than with the questionnaires distributed at the end of the course. Many appreciated the timing of the feedback session, seeing it as an opportunity to suggest improvements that they would benefit from.

Instructors have also received the method positively. Many feel that this method of feedback provides their students with an opportunity to honestly say what aspects of the class they are unhappy with, and to suggest how the class could be changed to help them.

17.3.1 A Procedure for Conducting a Student Feedback Visit

Both authors were trained in the technique of collecting student feedback by personnel from the Center for Research and Teaching (CRLT) at the University of Michigan. The procedure presented here represents a slight adaptation of the method that CRLT instructional consultants routinely use to collect student feedback. Other references for information on this type of classroom visit are [30] and [84].

1. **Arranging The Visit.** Well in advance of the intended visit, contact the instructor to be visited, and set up a day and time for the visit. As with observational visits, confirm that the instructor doesn't have something like a test planned for that day. This is even more important for a student feedback meeting, as students may be overly focused on the test, or may simply get up and leave when the test is over.

   During the contact with the instructor, explain the process. Many instructors think that the students will just gripe, so hearing that you will be focusing and moderating the discussion may help alleviate these concerns. Make sure that the instructor knows that the feedback will be kept confidential.

   While explaining the process, make sure to cover the following five items.

   - Emphasize that the student feedback session will take between twenty and twenty-five minutes, and that the instructor should leave the room, including collecting papers and answering questions, with at least twenty minutes remaining in the class period.

   - Ask the instructor if there is anything about the class, or the way it is run, on which the instructor would specifically like feedback.

   - Ask instructor to acknowledge your presence in the class, or introduce you and the purpose of the visit to the class.

   - Set up a date, time and place to give the students' feedback to the instructor. You should both set aside an hour for this follow-up meeting.

   - Ask the instructor if there is an overhead projector available in the room. If so, take two blank overhead transparencies and an overhead marker to the feedback session.

2. **Conducting the Observation.** Arrive early for the class, and find a seat that will allow you to observe the instructor and the majority of students with ease.

   If you don't know the instructor, introduce yourself when he or she arrives, and confirm that the feedback session will begin between twenty and twenty-five minutes *before* the end of class.

   During the first part of the class, observe the class just as in an observational visit, taking any notes you feel are relevant. Recall that it is important to look at what the students do as well as what the instructor does.

3. **Facilitating the Student Feedback Session.** The instructor should leave with at least twenty minutes left in the lesson. Introduce yourself to the class, and give a brief description of what they are expected to do. It is not uncommon for students to attempt to get up and leave at any time after the instructor finishes their lesson, so immediately emphasize that this feedback session is meant to improve their experiences in the class.

   Form the class into groups of three or four students. If the students have been working in groups during the lesson, rearrange them to form new groups in order cut down on the amount of off-task activity. Ask each group to appoint a recorder, and distribute the response sheets. These sheets ask the students to **List the major strengths of this course. (What is helping you learn in the course?)** and **List changes that could be made in the course to assist you in learning.** Examples of these sheets can be found in Section 17.3.2. Encourage students to be as specific as they can, to write down examples whenever they can, and to try to suggest changes that are practical.

Give students five to seven minutes to work in their small groups. Circulate and monitor the groups' progress. Some groups may get off-task, and because of the limited time, it is important to get them back on-task as quickly as possible. It is helpful to make an announcement when there are one or two minutes left for discussion. While you are circulating, try to note at least one group that has listed positive feedback under the question **List the major strengths of this course. (What is helping you learn in the course?)**

If the classroom has an overhead projector, set it up while the students are finishing their discussions. Label the top of one of the blank transparencies **Strengths** and label the top of the other **Changes**. If there is no overhead available, give copies of the forms that simply list **Strengths** and **Changes** to two students, and ask them to transcribe whatever you write on the board. Make two columns on the board, and label them **Strengths** and **Changes**.

Call the class to order, and go around each group, asking them to contribute a strength. Begin with the group that you noticed had written at least one strength. If the first group fails to list a strength, this can change the tone of the session and inhibit other groups from sharing their listed strengths. Record the strengths verbatim on the overhead transparency, or on the chalkboard. If you are working on the chalkboard, ask the two students with the blank sheets to record verbatim what you write there. At each suggested strength, check for majority opinion with a quick show of hands. Only record those strengths that are agreed to by a majority of the class. Continue to go around the room listing strengths until the groups have exhausted their lists. Repeat the process with suggested changes for the class. Once again, take a show of hands on each change. This can be very useful in helping to develop priorities for the instructor. If someone suggests a strength or a change that needs clarifying, ask the student to be more specific or to give examples, so you can communicate their suggestions more accurately to the instructor.

If the instructor has asked for feedback on any particular issues, and no one has mentioned these issues, ask the class about them.

End the class on time, and collect *all* of the sheets that the small groups have filled out, and if applicable, the two summary sheets. Check the accuracy of the records, and erase the boards before the next class enters the room.

4. **Writing the Report and Preparing for the Meeting.** Type student comments from the two final forms or overheads onto two separate pages, one for **Strengths** and one for **Changes**. It is helpful to the instructor if the items are listed from most significant to least significant. If some student comments or suggestions are, in your professional judgment, unnecessary or inappropriate, there is no need to pass them along to the instructor. Sample reports generated from student feedback can be found in Section 17.3.3.

After typing the sheets detailing the students' responses, you will have an overall view of what the students were saying. Consult the observational notes from the classroom visit and the sheets collected from the groups of students to find specific examples to substantiate issues that the students raised. It is often helpful for instructors to focus on two or three key issues, rather than to try to address lots of small issues. Pick out two or three issues to emphasize when you meet with the instructor. If the instructor has asked you to focus on any particular issues, make sure that you are able to address these as well.

5. **Meeting with the Instructor.** Meet with the instructor at a mutually convenient time and place to discuss the class observed. Meet with the instructor while the class session is still fresh in his or her memory, and before the next class session if possible. When selecting a venue for the meeting, the same considerations used in the follow-up meetings to observational visits should be taken into account. Thank the instructor for allowing the classroom visit to take place, even if he or she had no choice in the matter. It can be helpful to acknowledge that soliciting student opinions on their

classroom experience can be intimidating and stressful. Try to get the instructor to talk about how that particular class went, or how the course is going in general. A typical comment will be that the visited class was unusual or deficient because time was lost to the feedback session. If the instructor raises this objection, ask the instructor how that class would have gone differently if all the normal time had been available. Other ways to get the conversation started include asking the instructor to articulate the goals for that session and how class time was used to try to achieve them, and asking the instructor about his or her concerns for the class or for the course.

When you feel it is appropriate, share the report with the instructor. Give the instructor the sheets one at a time, starting with **Strengths**. Otherwise, the typical pattern will be for the instructor to ignore the **Strengths** sheet, and concentrate exclusively on the **Changes**.

Give the instructor enough time to read over each sheet, and be ready to clarify points, especially on the **Changes** sheet. Draw on the specific student examples on the group sheets and your observations from class to help clarify the overall comments. Ask the instructor to talk about his or her reaction to the students' comments on each sheet. Have the instructor talk about how the students' comments fit into the goals of the course, and what changes he or she may be able or willing to make. Attempt to focus the instructor on the two or three most important issues. Sometimes an instructor will have very good reasons *not* to make changes suggested by students. In this case, the instructor can be encouraged to share (with the class) his or her reasons for *not* making the changes.

Help the instructor plan a response to the class. Stress that it is very important to address the students soon after the feedback session. When offered the opportunity to express their opinions, the students will likely feel that they have been given input into the instructional climate of the class. If a response from the instructor is not forthcoming, they may feel cheated and resentful about the wasted opportunity. Help the instructor plan how to explain the implementation of changes, and how to articulate his or her reasons in the situations where the suggested changes are unreasonable or not in accordance with the course goals. The instructor should not feel the need to address every issue, only the most important ones. The entire response need not take more than five to ten minutes of class time.

6. **Following Up.** Check back with the instructor after a couple of weeks to see how the class is going, and how the changes that they have made have affected the learning environment. Some instructors will not relish the prospect of a follow-up, but resistance that you encounter in some cases should be offset by the instances where an instructor needs and welcomes further input and advice.

17.3.2 Examples of Forms Used for Student Feedback Visits

The examples given in this section are adapted from forms routinely used by instructional consultants at the University of Michigan to collect student feedback.

Midterm Student Feedback Form

In your small group, please discuss the following categories (strengths and suggested changes) and come to a consensus in the group on what should be written for the instructor. Being explicit and using good examples will make your comments more useful for the instructor. Please have a recorder write down the comments that you all agree on.

I. List the major strengths of the course. (What is helping you learn in the course?) Please explain briefly or give an example for each strength.

        <u>Strengths</u>   <u>Explanation/Example</u>

1.

2.

3.

4.

5.

II. List changes that could be made in the course to assist you in learning. Please explain how suggested changes could be made.

        <u>Changes</u>   <u>Ways your instructor can make changes</u>

1.

2.

3.

4.

5.

17.3.3 Examples of Information Collected and Reports from Student Feedback Visits.

1. **List the major strengths in the course. (What is helping you learn in the course?) Please explain briefly or give an example for each strength.**

Strengths	Explanation/Example
1. He gives us good handouts of how to do certain problems in the book.	—
2. He knows what he's talking about and goes beyond the book.	—
3. We know he wants to help us but he can't put his thoughts to words in english.	→always says see me in office hrs.
4. Clear about homework, assignments and quizes.	
5.	

II. **List changes that could be made in the course to assist you in learning. Please explain how suggested changes could be made.**

Changes	Ways your instructor can make changes
1. He doesn't go step by step, He assumes we know it all and only confuses us more.	Explain things slowly and in steps, make sure everyone understands before we go on.
2. He never really answers or explains our questions, he just goes over the same example.	Listen to us more, or get us to come up to the board
3. He gets frustrated and tells us its easy and only pre-cal. This makes us feel dumb	be nice + considerate.
4.	

Strengths
- Knows what he is talking about. (90%)
- His willingness to recommend us to office hours (100%)
- He writes explanations on the board. (30%)
- Makes sure we know what the assignment is. (60%)
- Handouts give good explanations. (100%)
- Responsive to questions from class. (75%)
- Willingness to do difficult Home work problems. (100%)

Changes
- Explain concepts more. (100%)
- Don't assume that we understand the problem. (100%)
- Don't skip steps. (100%)
- Follow the book in a more orderly manner. (90%)
- Explain problems in more than 1 way. (75%)
- More class participation. People working at board.
  In class group work. (50%)
- Don't get frustrated with questions. (100%)
- More control over class (90%)
- Routine schedule for class — HW, New Section, question (90%)
- Less technical explanations (100%)

Instructor :
Class :
Date :
Observer :

What is helping you learn in the course?

- <NAME> knows what he is talking about

  - Very knowledgeable about math.
  - He knows what he is talking about and goes beyond the book.

- <NAME> is eager to help us outside class– recommends office hours

  - He is always willing to see us at his office for help.
  - We know he wants to help us → always says see me in office hours.

- He makes sure that we know what the assignments are

  - His communication about assignments — he always writes the assignments on the board.
  - Clear about homework, assignments and quizzes

- Handouts give good explanations

  - The handouts are great
  - The example problems which are handed out are very helpful to us
  - Worksheets/handouts — explain problems
  - He gives us good handouts of how to do certain problems in the book

- Group homework

  - Group homework has helped us understand the material better. We are able to explain the material better from different points of view

- <NAME> is responsive to questions

  - Very responsive to questions.

- <NAME> gives help with individual homework problems

  - If the students have trouble with a specific problem, the instructor will explain it for us
  - <NAME> does many of the homework problems

- Quizzes are good

  - Quizzes are a good test to see if we are grasping the material
  - These require us to review the material, understand the material, give us an idea of what to expect for a grade

Changes to assist you in learning

- Explain concepts more (100% agree)
- Don't assume that we understand the problem (100% agree)

    - After a couple of steps into the problem, ask us if we understand.
    - Don't skip steps when we don't understand — we get even more confused.

- Don't get frustrated with questions (100% agree)

    - He gets frustrated and tells us its easy.

- Slow down (100% agree)

    - Go slower and take the math down to our level — explain problems step by step.
    - Slow down. Take more time working through examples.
    - Slow down — explain the steps in more detail.
    - Explain things slowly and in steps. Make sure that everyone understands before we move on.

- Give less technical explanations (100% agree)
- Follow the sequence of topics in the book (90% agree)

    - Follow the methods in the book. Use the book not only for examples but also for discussion.

- More control over the class (90% agree)
- Adopt a routine schedule for class (90% agree)

    - e.g., Individual homework, then the new section, then questions.

- Explain problems in more than one way (75% agree)

    - He sometimes doesn't answer our questions — he just goes over the same example. Listen to us or get us to come up to the board.

- More class participation (75% agree)

    - More in-class group work.
    - More people working at the board.
    - Have us do more work on the board or in groups to get us to think.

17.3.4   Advantages and Disadvantages of Student Feedback Visits

Advantages of Student Feedback Visits

- The feedback comes straight from the students.

- This type of visit allows the students to have input during the semester, when it can potentially improve their own experiences in the course.

- Students often pick up on things of which outside observers are completely unaware.

- Since the items under discussion are open-ended, the instructor can find out what is really most important to the students.

- The consensus building aspect of the process removes isolated, extreme opinions.

- If done correctly, this kind of activity can improve student morale and satisfaction with the course.

Disadvantages of Student Feedback Visits

- Some instructors may not take the responses seriously unless you can indicate *precisely* what the students were talking about, and so you must take care to find out exactly what students have in mind when they describe changes.

- If you are the only person visiting a fairly large number of classes, then you may be overwhelmed by the sheer number of classes to visit, the number of complaints that some classes generate, and the rudeness that some students may display either to their instructor or directly to you.

- Instructors may be nervous at the prospect of receiving feedback from their students, and may resent feeling "evaluated" by their students. This may lead them to act in a hostile or detached manner.

- Students may make unrealistic or impossible suggestions for change, and including these in the report undermines the credibility of the suggestions over which the instructor has some measure of control.

- Some students may give non-constructive feedback, which may be construed as whining.

- Some students may feel that they have been given license to continue to complain to the instructor about any aspect of the class about which they are dissatisfied.

- The students often have high expectations that their instructor will address their suggestions in some way. If the instructor simply chooses to ignore the feedback, the morale of the class can worsen.

- Some instructors consider student feedback, or at least its collection, to be a waste of time.

17.4   Collecting Student Feedback Without a Visit

Circumstances may arise where an instructor wants to collect student feedback, but it is impossible or impractical for an observer to visit the class and facilitate a student feedback session as described in Section 17.3.1. Under such circumstances, it is feasible for the instructor to collect feedback themselves using a similar process. The procedure described in this section shares many of the advantages and disadvantages of the procedure from Section 17.3.1, and the advantages and disadvantages particular to this process are spelled out at the end of this section. This process is inferior in its potential impact, although it is undoubtedly more convenient to arrange and implement.

17.4.1   A Procedure for Collecting Your Own Student Feedback

1. Divide the class up into groups of three to four students. Ensure that these groups are not groups that normally work together in class. Have each group appoint a recorder.

2. Hand each group a sheet for recording their comments and suggestions, and then leave the classroom. Sample sheets are found in Section 17.3.2.

3. Each group should spend five to seven minutes discussing both what is going well in class, and producing suggestions for change. The groups should be as specific as they possibly can, and provide examples illustrating their points.

4. As the groups come up with ideas, the reporter will record those with which the group, as a whole, agrees.

5. At the end of the class period, a student will collect the papers from each group and arrange for them to be returned to the instructor.

Instructor's Tasks

1. Set aside 15 minutes at the end of a lesson for the activity.

2. Explain the spirit and intention of the feedback exercise. This could include the following points.

   - This gives people in the class a chance to voice their opinions and suggest changes now, while they will benefit from the changes.
   - The procedure is anonymous and confidential.
   - Positive comments and suggestions for change are both requested.
   - Responses should reflect a group consensus.
   - The instructor values the students' opinions, and will take responses seriously.
   - The instructor will consider suggestions for change and make a genuine attempt to address the students' concerns.
   - The students should concentrate on those aspects of the course over which the instructor has some amount of control.

3. Describe the process to the students.

4. Appoint a student to collect each group's paper and return them to the instructor. You may either ask the student to put the papers in an envelope and bring them to your office, or to put the papers in an envelope with your name and departmental address written on it, and deliver the envelope to your department's office. If you choose the latter option, make sure to arrange to obtain the envelope from the departmental support staff. Do not simply ask the students to leave the papers in the classroom, because this compromises the confidentiality of the feedback, and increases the risk that the papers will be lost or discarded.

5. Tell the students that the process of reflection and planning will take a few days, but that you will respond to the suggestions.

6. Leave the room, and don't just wait outside the door.

7. When you have collected, read and reflected upon the students' responses, try to formulate a plan for responding to the suggestions. Five basic elements which your plan could include are

   - a statement acknowledging the students' feedback,
   - ideas for implementing the desired changes,
   - ideas for dealing with impossible requests (e.g., "No team homework"),

- ideas for how you will make students aware of your efforts to implement changes, and

- ideas for how you will establish whether or not the changes have been effective.

8. Meet with another instructor to discuss whether or not the feedback process has been useful to you, and any suggestions you have for improving the process. This is also an opportunity to discuss the responses from your students, especially any you found disturbing or surprising, and your response plan.

9. When you address the students' suggestions in class, begin by thanking them for their contributions, and for bringing so many valid issues to your attention. Try to avoid being defensive. If you decide that some of the suggestions for change are impractical, indicate your reasoning rather than just ignoring the suggestions. Tell them that you will concentrate on the issues that you believe you can change most effectively. Try to leave the students with the impression that you have taken their suggestions seriously.

17.4.2 Advantages and Disadvantages of Collecting Your Own Student Feedback

Advantages of Collecting Your Own Student Feedback

- This type of feedback collection is easier to arrange than a student feedback visit, and does not require an outside facilitator.

- Some instructors who resist having an observer visit their class still have an opportunity to collect feedback using this procedure.

- This procedure usually takes less class time than a student feedback visit.

Disadvantages of Collecting Your Own Student Feedback

- The feedback obtained via this method can be less rich and less useful than the feedback obtained when there is an observer present to clarify and expand on the meaning of students' suggestions.

- There is no filtration of inappropriate, rude and impossible suggestions.

- It can be difficult to find priorities in students' suggestions when there is no observer collecting feedback using a procedure that aims to find consensus among the students.

- If the instructor does not talk about the feedback with anyone else, then there is the possibility for him or her to take students' comments very personally. It is possible for the instructor to be completely turned off teaching by inappropriate student comments.

17.5 Peer Visits

17.5.1 A Procedure for Implementing a Program of Peer Visits

A less labor-intensive and possibly less threatening method of generating observations and feedback for instructors is to arrange a system of peer visits. After the instructors have taught for long enough to develop a sense of what they feel does or does not work well in the classroom, pair the instructors, and ask them to conduct observational visits for each other. The visits should be scheduled close together, and the instructors should meet to share feedback as soon after the second visitation as possible.

To prepare the instructors for these visits, explain the goals of the peer observation process. Give them a handout stating these goals, which you can modify from the goals found in Section 17.1.2. As further preparation, you should give them guidelines on conducting the process and suggestions concerning basic observational and feedback skills. You can break down the guidelines using the categories found in Section 17.2.1:

1. setting up,

2. the visit,

3. writing the report and planning for the meeting, and

4. the meeting.

You can use the advice given in that section and your own experiences to formulate the guidelines that you give to the instructors.

Decide on some mechanism for feedback from the peer visit program. After the pairs have completed their peer observations, ask them to submit written reports, or conduct an informal round-table exchange on what they learned and observed.

17.5.2   Advantages and Disadvantages of a Program of Peer Visits.

Advantages of Peer Visits.

- A program of peer visits can lend a collaborative spirit to your department, and give individual instructors added peer support.

- This type of visit requires less time and energy from the course coordinators. The only cost involved is matching the instructors with a partner with whom they will exchange visits.

- The instructors may be less fearful of this type of visit, since the observer will not be someone in a position of authority. In addition, the observed will become the observer and vice-versa, taking away any feeling of hierarchy between the instructors.

- It is a common experience for course coordinators to gain valuable teaching insights from observing classes and formulating feedback. Peer visits give instructors exposure to alternative teaching strategies, and provide them with a chance to develop observation and feedback skills, which they can later use to critique their own teaching.

Disadvantages of Peer Visits.

- Because a course coordinator is not directly involved in these visits, some instructors may shirk their responsibility to do the peer visits.

- Because the observers are likely to be inexperienced instructors, there is the possibility that they will not be able to provide constructive feedback to the instructors whom they observe.

- Some instructors may feel a sense of competition, and therefore focus exclusively on the perceived shortcomings of the instructors whom they visit.

17.6   Potential Problems with Implementing Class Visits

17.6.1   Instructor Cooperation

There may be cases of significant resistance to a program of class visits. Many instructors feel that their classroom is their personal space, and they are uncomfortable at the prospect of allowing outsiders in to comment on what is occurring there. They may also be quite resistant to the idea of allowing students an opportunity to voice their opinions on the teaching at a time when the instructors and students still have to face each other. In addition, teaching takes a significant amount of time and energy, especially when combined with a full-time research or class work program. Because of this, some instructors may feel as though the time involved in gathering and dealing with feedback is a burden instead of an opportunity for growth. Finally, if the class visits are used for evaluative purposes, especially hiring or promotion decisions, many instructors will naturally feel threatened by the pressure to perform well.

To encourage instructor cooperation, consider the following suggestions concerning a program of class visits.

- Keep the process entirely confidential, and clearly communicate this fact to the instructors.

- Especially for beginning instructors, use the class visits for assessment not evaluation. If the instructors understand that the purpose of the visits is to help them improve their teaching practice, and not to threaten them, then much of their defensiveness will be removed.

- Give constructive, compassionate, credible, and concrete feedback. Concentrate on some things that are going right in the class, and some areas of deficiency where improvement is within reach. Acknowledge that teaching is a complex and difficult job, and that it takes time to master. Use your experience as a mathematics instructor and instructor trainer to speak to their mathematical and pedagogical sensibilities. Finally, concentrate on concrete suggestions that the instructors can put into practice for immediate and tangible benefits.

- Be efficient in running the program, and in so doing, communicate that you understand that the instructor's time is valuable.

- In spite of your best efforts, some instructors may try to avoid visits by not responding to your calls or E-mail messages. If you are serious about visiting classes, you must be persistent in your attempts to contact these instructors.

17.6.2   Student Cooperation

During student feedback sessions, the students may be resistant to the process, or even contemptuous of the facilitator. The following suggestions are meant to alleviate some of the student resistance.

- It helps if the instructor tells the students before leaving the classroom that the students' feedback is welcomed and valued.

- Clearly communicate to the students that they have an opportunity to favorably affect changes in the class, at a point in time when it will still benefit them.

- Communicate to the students that the patterns of feedback collected may be used to improve the overall nature of the teaching in the department.

- Be polite, in control, and extremely efficient. Explain the procedure clearly, and conduct it confidently.

17.6.3   Before The Visits

Although classroom visits are designed to benefit everyone involved in the course, some instructors and students display an amazing amount of contempt and cynicism. The following is a partial list of unpleasant situations that may occur. Before conducting observational visits or student feedback visits, the facilitators should reflect on what they will do if any of these situations arise.

- An instructor ignores your attempts to contact them to set up a class visit.

- An instructor sends the following message in response to your attempt to set up a visit.

  > Hello. Thank you for your message. I do not think that it will be necessary for you to visit my class. I am well aware of the areas of my teaching that I have to work on, and have enough to keep me occupied for several years. I doubt that this visit will do anything but confirm what I already know. I also feel that this visit will be an unnecessary obstacle to staying abreast of the curriculum. As you well know, there is a great deal to teach in this course. I think that class time is valuable, and should be spent in ways that benefit the students, i.e., teaching the subject matter of the course.

- An instructor agrees on a date for a class visit. The evening before, you receive a message from the instructor asking to postpone the visit, because the instructor has decided at the last minute to give a test.

- You set up a visit, and insist that the instructor *not* give a test on the day of the visit. Nevertheless, when class begins, the instructor asks the students to put away their notes, and hands out a test.

- The instructor selects a day for a student feedback visit that is devoted to exam review. When the instructor announces that the review is over with twenty–five minutes of the lesson remaining, some students become openly angry that their questions were not addressed.

- Despite confirming that you need at least twenty minutes at the end to collect student feedback, the instructor just keeps going on with the lesson past this time.

- When the instructor leaves to allow you to collect feedback, several students pack up their belongings and attempt to leave.

- You provide the students with instructions on how you want them to break up into small groups to discuss what had been going on in the class; however, no one budges, and several students complain about being required to move.

- You have formed the students into groups for a student–feedback visit, asked them to appoint a recorder, and to discuss strengths and changes. Several students get up from their groups, and start conversations elsewhere that have nothing to do with the strengths and changes.

- When you ask the small groups of students to pay attention to the board and contribute their strengths and changes, two students in the class attempt to take over by pressing their own agendas and problems with the class into every groups' comments.

- One group of students suggests a change that amounts to a personal attack on the instructor.

- When you recover the two summary sheets from a student feedback visit, you discover that they have different comments recorded on them, some of which you don't remember writing up on the board.

- You go to the meeting that you have arranged with the instructor to give them your own or the students' feedback, but the instructor fails to appear.

- The instructor meets with you to receive feedback, but he or she is defensive and completely resistant to honest reflection. According to the instructor, and contrary to your observations, nothing about the class could possibly be improved.

- You arrange a system of peer visits, but many of the pairs fail to execute their visits.

- During an exchange of peer visits, two instructors become competitive, which leads to open hostility.

17.7   Conclusion

Although there can be difficulties involved in implementing a program of class visits and feedback, most instructors find the process very beneficial to their professional growth. Aside from the time and energy involved for the trainers, the main item of difficulty involved with classroom visits is the occasional resistance by instructors engendered by their unnecessary fear of being "evaluated." We reiterate that to minimize this apprehension, the trainers should use visits for the professional growth of their trainees and not for evaluation purposes.

Because a schedule of visits and feedback affords the most personalized support to instructors involved in an integrated professional development program, and because it can be very worthwhile to the trainers as well, it is a most vital aspect of the program that should not be avoided or minimized because of the time and effort involved. Finally, to obtain the maximum benefit from a schedule of visits and feedback, the trainers should plan to do them early in the semester, and they should be well prepared, supportive, credible, and thorough in their feedback.

17.8   Suggested Reading

This chapter on classroom visits has addressed a number of issues such as the practicalities of visiting classes, providing feedback to instructors and some of the program-wide uses of the information that can be gathered from a program of classroom observations. In addition, there are methods of observing classes that are in wide-spread usage (e.g., videotape, peer visits) that we have described briefly or not at all.

As might be anticipated, a wealth of literature exists on each of these topics. We offer a few (we believe) representative examples of this literature here. Due to the extensive list, we have broken down the suggested readings into the following sections:

1. Checklists for Conducting Classroom Observations.

2. Procedures for Making Observations and Collecting Student Feedback.

3. Videotape and Peer Visits as Tools for Improving Teaching.

1. **Checklists for Conducting Classroom Observations**

- Maryellen Weimer, Joan Parrett and Mary-Margaret Kerns. *"How Am I Teaching? Forms and Activities for Acquiring Instructional Input."* [119]

  This book contains a very extensive and detailed checklist intended for use by classroom observers. It features an impressive array of questions intended for classroom observers to rate the instructor, students and general features of the class against. These questions are arranged into the following categories: importance and suitability of content, organization of content, presentation style, establishing and maintaining contact with students, and questioning ability.

- Maryellen Weimer. *"Improving College Teaching."* [118]

  This book does not feature an extensive checklist like the earlier reference of Weimer, Parrett and Kerns. However, it does feature a chapter titled, "Ongoing Assessment and Feedback" (Chapter 4) that discusses the formulation and use of these kinds of instruments in considerable detail. (The chapter also discusses other approaches to assessing teaching besides the development and use of checklists.) The discussion of ongoing assessment is situated within a five step plan for instructional improvement: develop instructional awareness, gather information, decide what to change and how to change it, implement changes, and lastly, assess effectiveness.

- Sharon Baiocco and Jamie DeWaters. *"Successful College Teaching. Problem–Solving Strategies of Distinguished Professors."* [6]

  Appendix D of this book ("Tools for Developing Professionalism") lists a number of rubrics that may be readily adapted into guidelines for what to look for when conducting classroom evaluations. The four rubrics listed are: planning and organization, classroom management, methods and techniques, and personal. Listed beneath each rubric is a list of criteria that, in the opinion of Baiocco and DeWaters, reflect a high standard of professional conduct for university and college faculty to aspire to in regards to teaching. The main difficulty that we see with this approach is that the specification of these professional standards is given in extremely decontextualized terms. For example, under "Methods and Techniques," one standard is: "Use of the board, overhead, and corresponding seat work was very effective." Effective use of the "tools of the trade" of college teaching is certainly something that one would expect to see in high–quality teaching, but it is not very suggestive of just what "effective use" actually is, and what an observer might look for as evidence of this.

2. **Procedures for Making Observations and Collecting Student Feedback**

- Mark Redmond and D. Joseph Clark. "Small group instructional diagnosis: Final report." [92]

  This technical report describes a procedure for systematically collecting student feedback that was developed at the University of Washington. The process is called "Small Group Instructional Diagnosis" (SGID). The report describes the procedure that was used to collect student feedback (similar to the process of "Midsemester Feedback" described in this chapter) and reports positive students' responses to the use of SGID in a large number of classes. A small deficiency of this report is that it refers to several appendices that are not included in the version available from the ERIC Document Reproduction Service.

- Mark Redmond. "A process of midterm evaluation incorporating small group discussion of a course and its effect on student motivation." [91]

  This report is related to the technical report of Redmond and Clark (1982) listed here. The report also describes the SGID procedure, but reports on a study of the effects of performing SGID on the levels of motivation of students in the course. Using a questionnaire with a group of 199 students (107 who had experienced SGID and a control group of 92 who had not experienced SGID) the study found significant improvements in some areas of motivation for the SGID students, as compared with the control group. The author suggests that SGID may be used as an alternative to end-of-semester questionnaires.

- Charles Mitsakos. "Mirrors for the Classroom: A Guide to Observations Techniques for Teachers and Supervisors." [83]

  This is a fairly extensive (47 pages) guide intended to provide classroom observers with some idea of useful things to look for, and concrete techniques that can be employed to observe specific phenomena. The specific phenomena that the techniques are designed to observe fall into four categories: the amount of time students spend working on learning tasks, the amount of teacher talking that goes on, the amount and quality of student–teacher and student–student interaction in the class, and the teacher's classroom management skills. The guide suggests that observers consult with the teacher before observations are made, and even suggests that the specific observational techniques to be employed be negotiated during this initial meeting.

- Mary Sorcinelli. *"Evaluation of Teaching Handbook."* [107]

  In addition to a great deal of other resources for evaluating teaching at the university level, this handbook contains a section on classroom observations. This reference is remarkable because of the amount of detail it devotes to the pre-visit conference between observer and instructor. Sorcinelli suggests eight questions that observers can use to draw instructors out, and develop a clear picture of what the instructor intends to happen in the class period that will be observed, how that is related to what occurred in recently taught classes, and if there is anything that the instructor would particularly like feedback on. The author gives a number of rubrics for things that observers might look for, although we note that these are tied to *a* particular model of teaching excellence that may or may not apply to your institution. These rubrics are: knowledge of the subject matter, organization and clarity, instructor—student interaction (including the use of questions), presentation and enthusiasm, and student behavior. Sorcinelli also includes a list of questions that the observer may use in the post-observation interview to draw the instructor out, and get him or her talking about what went on in the class that was observed. Note: this reference is not easy to obtain, although it has been reprinted in the more widely available book:

- Sara Jane Coffman. "Improving your teaching through small-group diagnosis." [30]

  This article describes the use of a student feedback procedure based on the SGID process. The author briefly describes how student feedback is collected and used at Purdue University. In this approach, the use of consensus-building is very strongly emphasized. The author also describes the reactions of faculty to the student feedback, and reports that this is generally very positive. Some faculty members are reported to have said that the information generated by this method was much more applicable to improving their teaching than the comments and ratings customarily collected through end-of-semester course questionnaires.

- Carolyn Frank. *"Ethnographic Eyes: A Teacher's Guide to Classroom Observation."* [52]

  In this book, the author suggests using methods of observation from anthropological and social research (principally ethnography) to develop "lenses for seeing the patterns and practices of life within classrooms." The author's goal in this book is to develop techniques of observation that enable teachers to "read between the lines" of classroom events, and begin to probe what lies beneath the surface of the patterns and events that take place in class. Perhaps because the book is an application of the methods of ethnography to classrooms, many of the suggestions for how to record and analyze classroom observations have as their goal the description of the "culture" that is created as students and teacher interact in the classroom. As such, many of the methods assume that the program of observations will be very in-depth and ongoing, which may make them less useful to mathematics faculty who are trying to help large numbers of novice instructors to improve or refine their teaching methods. However, this book would certainly be a valuable reference for individuals seriously interested in improving a single course, and who have the time, energy and expertise required to carry out an intensive program of observations in a single classroom.

- Christopher Knapper and Sergio Piccinin, eds. *"Using Consultants to Improve Teaching.* New Directions in Teaching and Learning #79." [68]

  This book is part of the "New Directions in Teaching and Learning" series, and features a number of articles by noted scholars of teaching. Many of the articles discuss teaching consultation in fairly broad terms, and are not of direct relevance to classroom observation. Likewise, the approaches to teaching consultation often involve much more than just classroom observations. One of the articles, "How Individual Consultation Affects Teaching" describes some of the positive effects that providing feedback to faculty on their teaching may have. The final chapter, "Resources on Instructional Consultation," offers a very extensive list of the materials that can be useful in consultation, as well as an extensive list of institutions with expertise and experience in consultation.

- Patricia Shure. "Early Student Feedback" in MAA Notes #49. [55]

  This article briefly describes the use of early semester feedback from students in the introductory courses (predominantly first and second semester calculus) at the University of Michigan. The article describes the goals of these introductory courses in detail, and gives some indication of the ways in which the feedback is used to review individual instructor's progress towards implementing these goals. The description of procedures for collecting and interpreting the student feedback is extremely brief, but is essentially similar to the "Mid-semester Feedback" process described here. This article also describes some ways in which students' feedback may be interpreted, rather than taken at face value.

- Susan Sullivan and Jeffrey Glanz. *"Supervision that Improves Teaching: Strategies and Techniques."* [110]

  This book deals broadly with the supervision of instructors at all levels of education, including higher education. The scope of this book is much broader than conducting classroom observations, but it does devote a chapter to this topic. Chapter 3 of this book describes fourteen tools for conducting classroom observations that have been used successfully in college classrooms, as well as in school settings. Sullivan and Glanz also describe an integrated approach to instructional supervision, and in other parts of the book describe ways in which information from programs of classroom observation and feedback may be integrated into an overall supervision plan.

3. **Videotape as a Tool for Improving Teaching.**

- Barbara Davis. *"Tools for Teaching."* [36]

  Chapter 42 of this extensive collection of advice and techniques for college teaching discusses the use of videotape to improve instruction. Unlike other references listed here, the emphasis in this chapter is on instructor self-evaluation, rather than on meeting with an instructional consultant to watch and analyze the tape. Davis outlines a very clear procedure for recording and analyzing one of your own classes, including a list of steps to do to set up the taping, and a very extensive and explicit process for viewing and analyzing the tape after the class. Davis suggests that after the class, instructors should: view the tape as soon as possible, plan to spend about twice as long analyzing the tape as it took to tape the class, view the tape with a supportive consultant or colleague (if one is available), and first go for a "big picture" rather than details. Davis also suggests viewing the tape again to concentrate on specific areas of interest or concern, such as the frequency and type of interactions between members of the class, the number of types of questions, voice characteristics or body language. Finally, Davis gives a checklist to focus attention on specific areas of the teaching performance, such as organization, style, clarity, questioning skills, student interest and classroom climate.

- Joel Hamkins. "Using video and peer feedback to improve teaching" in MAA Notes #49. [55]

  This article very briefly describes a program of graduate student instructor (GSI) development at the University of California, Berkeley. The article describes the integration of both videotaping GSIs teaching their classes, instituting a program of peer visits (in which GSIs visit each others classes to conduct direct observations of teaching without the use of videotape) and a series of meetings in which GSIs discuss issues related to teaching. The author gives a handful of suggestions on running such a program, and recommends that each GSI be taped twice during the course of a semester. The author feels that the first taping often reveals little about the teaching, as the GSI is often very self-conscious, and that the second taping reveals a great deal more.

- Deborah Bergstrand. "Exchanging class visits: Improving teaching for both junior and senior faculty" in MAA Notes #49. [55]

- Pao-sheng Hsu. "Peer review of teaching" in MAA Notes #49. [55]

  These two articles (both very short) describe programs of "peer visits" in which faculty members visit each other's classes and offer suggestions and feedback. In the second article (by Hsu), the author invites teams of observers — consisting of mathematicians and faculty from other disciplines — to observe her class.

Appendix A

Tips for Running Meetings

The chapters in this book provide detailed information for orchestrating training meetings. Trainers are frequently faced with situations that the meeting descriptions do not anticipate. This appendix is included as a resource for dealing with the inevitable surprises and problems that arise when running effective training meetings. This is a helpful section to read *before* running any training meetings at all. It is also intended to serve as a reference for dealing with recurrent problems. This appendix is intended to be helpful, but not encyclopedic.

A.1 To Meet or Not To Meet

When asked how to improve meetings, many people will supply one or more of the following suggestions.

- Make the meetings shorter.

- Have fewer of them.

- Make the meetings more relevant to me.

- Get rid of them altogether.

The unfortunate reality of the situation is that many people have no love for meetings, and many consider meetings to be a complete waste of time. Part of this stems from the fact that many meeting participants have difficulty organizing their time and priorities, but some also stems from the fact that meetings can be long, boring, and unnecessary. Before planning or announcing a meeting, trainers must ask themselves the important question,

"Do we really need to have a meeting at all?"

If the honest answer to this question is "No," in other words, if there is an equally effective and reliable way to convey information to instructors or help them make sense of their experiences, then trainers should seriously consider *not* holding a meeting at all. Rather than a meeting being a foregone conclusion, the meeting should be the result of *consideration* and *decision*.

Meeting Objectives

In order to determine whether a meeting is really required or not, it is suggested that trainers develop objectives or goals for the instructors involved in their training program. Examples of objectives may include some of the following.

- To help instructors become proficient at grading homework, so that they are able to grade quickly, and achieve high-quality results.

- To help instructors develop a well-defined idea of how they will use quizzes and other in-class assessment in their overall grading scheme.

- To find out what difficult classroom situations instructors have faced, and what they have done about them.

- To make sure that all sections of the syllabus will have been taught before the mid-term exam.

All of these objectives could be addressed in large-scale meetings of all the instructors involved in the course, but they do not *have* to be. For example, E-mail may be a less intrusive, more convenient way to find out if all sections of the syllabus will have been taught before the exam. Other objectives could be realized through individual interviews or meetings with small groups of instructors.

As a final remark, the point of this section is *not* to suggest that instructors should never meet. The purpose of this section is to suggest that many people react negatively to meetings, and alternatives to large meetings can often be utilized just as effectively to achieve the objectives of your training program.

A.2   What Makes Meetings Work?

Haynes [57] suggests a range of criteria that contribute to effective meetings. The list given below is based on this range of criteria. The criteria that are most relevant to the kinds of training meetings described in this book may be grouped into five categories: organization, environment, time management, participation, and responsibility. These are summarized below.

1. **Organization**

   - The trainers have developed clear objectives, and a meeting is the most effective way to achieve those objectives.
   - An agenda is devised and prepared *well in advance* of the meeting.
   - If appropriate, meeting participants have an opportunity to contribute to the meeting agenda.
   - The selected meeting time and day is one that all participants can make.
   - Participants receive notice of the meeting day, time, and place *well in advance*.
   - It is determined which agenda items can be treated lightly or omitted if important items take longer than expected to deal with.

2. **Environment**

   - The meeting room is large enough to comfortably accommodate all participants.
   - The meeting room can accommodate the activities planned for the meeting. For example, if the participants will be expected to work in small groups, the room should facilitate this.
   - The meeting room is not too hot or too cold. Some books recommend making a meeting room a little on the cool side, as warm rooms may encourage participants to nod off.
   - If the meeting is long, the participants will need access to facilities such as restrooms and refreshments.
   - Any needed audiovisual equipment is in good working order.

3. **Time Management**

   - *THE MEETING ENDS ON TIME!*
   - The meeting begins on time.
   - The meeting leaders monitor the use of time throughout the meeting, and try to stick as closely to the schedule as possible.

4. **Participation**

- If appropriate, everyone has a chance to present his or her point of view.
- When appropriate, there are periodic summaries of meeting progress.
- No one participant dominates discussions.
- If decisions are made at the meeting, everyone has a voice in the decision.

5. **Responsibility**

- Participants can be relied upon to carry out actions agreed to at meetings.
- Instructors can be counted on to attend meetings.
- The meeting leader follows up on questions that he or she was not able to answer during the meeting.

These criteria can serve as a starting point for examining the effectiveness of training meetings. Because institutions and teaching environments vary, meeting leaders may find that they need to customize this list for their particular institution. The modified handout may be circulated as a guide for meeting participants.

A.3 Addressing Complaints With Meetings

This section attempts to set out some of the ways that trainers may address common complaints with meetings. The suggestions presented here try to address problems through organizational and structural means. Suggestions for dealing with difficult meeting participants through interpersonal means can be found in Section A.6. The objective of this section is *not* to suggest that meetings *cannot* be effective, nor is it to suggest that meetings should not be held.

Haynes [57] quotes a survey in which 75% of executives were "bothered" by the ineffectiveness of meetings. Although business meetings typically have different objectives from the training meetings presented in this book, the fact that a large majority of people interviewed were negative about meetings is informative. Among college mathematics instructors, there is also a feeling that, although some meetings do provide useful information and help to impart skills, the payoff is not commensurate with the time and effort involved. The concerns of many college mathematics instructors over training meetings mirror the concerns of executives over business meetings. The seven most frequent complaints with meetings listed by Haynes are given below.

1. **Drifting off the subject**
2. **Poor preparation and planning**
3. **Questionable effectiveness**
4. **Lack of listening**
5. **Verbosity of participants**
6. **Length**
7. **Lack of participation**

Addressing these problems, if they occur, is a substantial part of running a training meeting that works well. The following suggestions are adapted from Haynes [57] to address the difficulties specifically mentioned above. Where possible, the suggestions listed below have already been provided in the descriptions of the training meetings in this book. For example, all meeting descriptions include suggested goals and agendas, and many have suggestions for managing meeting time. Other suggestions, for example taking charge of the situation, need to be implemented while the meeting is actually in progress.

- **Develop objectives or goals for the meeting.** Objectives, or goals, are necessary to decide whether or not to hold a meeting in the first place.

- **Only meet if absolutely necessary.** As indicated in Section A.1, if other possibilities exist that will allow the training program to achieve its objectives without a large, time-consuming meeting, then many instructors will appreciate it if these other possibilities are investigated and utilized. This is not to say that meetings should never be held, or that deciding to hold a meeting reflects a failure to find a viable alternative. The imperative here is to be respectful of participants' time and busy schedules, and to try to find convenient, effective alternatives to large meetings when possible.

- **State objectives or goals for the meeting.** If a meeting is called, make sure that everyone who will attend the meeting has a very clear idea of what the goals or objectives of this meeting are, and why they should attend. As mentioned before, meeting participants may be informed of the goals of the meeting beforehand, as well as at the start of the meeting.

- **Prepare an agenda.** The agenda is essential, as it tells participants what they can expect from the meeting, how the meeting time will be used, and what are appropriate forms and levels of participation. In this way, the agenda functions as a map for the meeting in that it can be used to guide wayward and meandering meeting participants back onto the appropriate path. It is essential to distribute the agenda to participants *well in advance of the meeting*, and to clearly indicate any preparations that meeting participants are expected to make.

- **If appropriate, be selective when picking participants.** Often, training meetings will simply be open to all. Some of the meeting descriptions in this book, for example the description in "Administrative Issues for the End of Semester," suggest that experienced instructors need not attend. Generally speaking, the fewer the number of participants, the greater the engagement and participation. On the other hand, more people usually means more perspectives, more experiences, and more ideas. Often, through class visits or conversations, it will become apparent that some instructors are struggling and will benefit substantially from certain meetings. For example, some beginning instructors have trouble controlling their class and could benefit from the "Establishing and Maintaining Control in Your Classroom" meeting. Under such circumstances it is appropriate to privately contact the instructor(s) concerned and encourage them to attend.

- **Manage meeting time.** All of the meeting descriptions in this book include detailed suggestions for time management, if the meeting is run as a formal training session. If the meeting leader decides on a less formal approach, a different approach to time management is also required. For example, Haynes [57] suggests "striking a balance between wasting time and railroading the group." Instructors who are familiar with active learning or cooperative learning probably have a well-developed sense of what is required here. On one hand, the meeting leader must allow sufficient time for participants to pair up or form groups, to read handout materials, and to think about discussion questions. On the other hand, the leader needs to *END THE MEETING ON TIME!*

- **Take charge of the meeting, but do so in your own way.** When holding a training meeting for college instructors, the amount of meeting time is usually in short supply. The patience of the meeting participants is often in short supply as well, not to mention that there will be little tolerance for off-topic discussions that prolong the meeting. As a result, effective control and guidance are required. The discussion must be kept on topic, and it is very important to *END THE MEETING ON TIME!* Sometimes, meeting participants may try to dominate the proceedings, or they may wander off the topic and begin discussing something only thinly related to the objectives of the meeting. In this situation someone must get the meeting back to the items on the agenda. It is unreasonable to expect that one leadership style will suit everyone who uses this book. However, it is important to point out the need for effective leadership in training meetings, and to encourage people who plan to run meetings to find a leadership style with which they are comfortable.

- **End with a summary.** This can be as simple as reiterating the meeting's goals and reviewing suggestions for change or action that have been advanced. Some meeting leaders like to end with a kind of "pep talk." If this approach is used, then it is important for the "pep talker" to do so with genuine sincerity, otherwise the gesture may be interpreted as a cheap ploy and do more harm than good.

A.4 Stimulating Discussion

A.4.1 Using Questions

This appendix encourages meeting leaders to plan for as much active participation and discussion as the constraints of meeting time will permit. Experience shows that it often falls to the leader to get activities, and particularly discussions, going. Some of the reasons that discussions can be hard to start are mentioned in Section A.6.3, and revolve around the assumption that the primary function of the group is to serve the individual.

Haynes [57] suggests that the meeting leader can use questions to encourage participants to join the discussion. Several of these ideas are summarized below, with examples of questions which can be used.

1. **Ask people for their opinions.** "How could you use that in your classroom?"

2. **Call on people by name.** "We've heard from everyone but Steve. What is your opinion on this, Steve?"

3. **Request summaries.** "Lynn just made some really good points. Darren, could you summarize them for us?"

4. **Clarify statements.** "I'm not sure I caught all of that. What would I do if . . . ?"

5. **Ask for Examples.** "Can you give me an example of when you'd do that?"

6. **Look for differences.** "Most people seem to agree on this. Is there anyone who sees the situation differently?"

7. **Explore details.** "That sounds really interesting, Cathy. How would you deal with . . . ?"

8. **Protect participants.** "Well, Wolfgang, you seem to have a lot to say against Leo's idea. What does Leo's idea have going for it?"

9. **Question assumptions.** "It seems to me that what you're saying would only work if everyone came up with an answer simultaneously. Do you think that this is possible or not?"

10. **Look ahead.** "Assuming that what you're saying will work, how could we use it for the difficult trigonometry sections next week?"

Haynes [57] also suggests things to avoid. Briefly, these are

- unanswerable questions;

- simple "yes" or "no" questions;

- vague questions—questions where the participants can't tell what the question is; and

- interrogations.

One last consideration is that a discussion is intended to be a *discussion*, not a series of conversations between the meeting leader and a few of the participants. If the meeting leader provides too much guidance and stimulation, then the so-called discussion can become a question and answer session, with only the meeting leader posing the questions.

A.4.2 The Physical Layout of the Room

Some rooms will encourage discussion, whereas others will not. Having all of the participants facing each other around tables may encourage discussion, whereas having the participants all facing the front of a lecture hall may not. Often, this is not something that the meeting leader will have control over, but if there are different rooms available for the meeting, then it is a factor that should be considered.

A.5   Roles for Meetings

Chang and Kehoe [28], among others, suggest that meetings are most effectively run when everyone involved in the meeting has a clear idea of the role that they are supposed to play. Two of these roles—leader and participant—are universally recognized, although specific areas of responsibility are usually less well appreciated. Two other useful roles are those of facilitator and recorder. In the remainder of this section, the specific areas of responsibility suggested for all who will take part in the meeting are listed. These lists have been adapted from those given by Chang and Kehoe [28].

In addition to setting out the areas of responsibility for the people who will be running the meeting, these lists may also be distributed to participants. Many participants are surprised to learn that they have a role to play in the meeting, beyond presenting themselves at the announced venue.

A.5.1   The Leader's Role

Most people recognize the need for two roles in a meeting situation: participant and leader. The meeting leader is principally concerned with *content*. The leader is, in some sense, the *owner* of the meeting. The leader decides what items will appear on the agenda, and how those items will be addressed during the meeting. The leader also has the power to stop what is going on in the meeting, if he or she feels that the meeting is getting away from the items on the agenda. This can be difficult when the meeting participants are people accustomed to working in an academic setting, with personal autonomy, and the freedom to pursue whatever interests them. This highlights the importance of circulating the agenda in advance of the meeting, and communicating the expectation that meeting participants will confine themselves to agenda items.

Preparing for the Meeting

The meeting leader

- develops objectives,

- prepares the agenda once objectives have been determined,

- schedules the meeting so that everyone who needs to attend can.

- (if appropriate) decides who needs to come to the meeting and contacts them,

- circulates the agenda and informs everyone of the meeting time and place,

- makes sure that everybody knows their role for the meeting,

- reads the plan for the meeting,

- prepares handout materials, refreshments, etc. for the meeting, and

- checks out the meeting room and any needed equipment before the meeting.

Conducting the Meeting

The meeting leader

- starts the meeting on time,

- states the goals or objectives for the meeting,

- establishes ground rules, if these are used,

- follows the agenda,

- helps to keep small groups focused on working productively,

- tries to keep the meeting running to the plan,

- retains the power to stop the meeting and change the format, especially if running out of time,

- summarizes key points, and

- ***ENDS THE MEETING ON TIME!***

A.5.2   The Participant's Role

This is the second universally recognized role for meetings, yet it is perhaps the least well understood. Unfortunately, some people think that participating in a meeting involves showing up to the meeting venue (not necessarily on time), enduring the long and usually boring speeches, and not pointing out that actual work could be done during the meeting time. Meeting participation is understood as a passive activity that takes up time that could be usefully spent. Potential meeting participants need to be convinced that this is exactly the *opposite* of what is expected. Circulating the following outline of responsibility to meeting participants prior to meetings is a step in the right direction. This assumes that the participants will take the time to actually read the outline. It may be necessary to pursue direct action in the case of particularly difficult individuals, in order to ensure that they realize what is expected of them as meeting participants.

Preparing for the Meeting

A meeting participant

- reviews the agenda before the meeting,

- confirms attendance at the meeting or contacts the leader *well ahead of the meeting* to make alternative arrangements, and

- does whatever preparation is required.

Conducting the Meeting

A meeting participant

- knows the objectives of the meeting ahead of time,

- arrives at the meeting on time,

- tries to keep an open mind,

- helps ensure participation from everyone at the meeting,

- sticks to the items on the agenda when engaged in discussion or group work,

- suggests alternatives when appropriate,

- shares helpful and useful ideas,

- supports and sticks to established ground rules, and

- does not create difficult situations (moreover tries to discourage others from doing so).

A.5.3   The Facilitator's Role

The meeting leader is concerned with the *content* of the meeting. The facilitator is concerned with the *process* of conducting the meeting. The facilitator acts as a time-keeper and referee for the meeting, trying to keep the meeting running according to the plan that has been devised, dealing with problem people, and remaining neutral in order to resolve any disputes that may arise. Many people assume that these are functions that the meeting leader naturally assumes. However, there are problems when the meeting leader is the one who mediates conflicts, as often the conflict will *involve* the meeting leader or the policies that he or she advocates. It is always easier if the person who mediates a dispute is *not* a party to the dispute.

Preparing for the Meeting

The facilitator

- develops objectives,

- reviews the agenda before the meeting,

- works with the leader on logistical problems,

- reads the plan for the meeting in order to have a good idea of how much time should be spent on each part of the meeting,

- does whatever preparation is required, and

- tries to identify and plan for difficult situations that may arise.

Conducting the Meeting

The facilitator

- focuses the participants on relevant issues,

- helps to stimulate discussion when appropriate,

- tries to get everyone to participate,

- monitors the amount of time spent on each agenda item,

- tries to keep the meeting running according to the plan,

- protects people and their ideas from attacks,

- deals with difficult people and situations,

- remains neutral if conflicts arise,

- tries to keep discussions focused on relevant topics, and

- tries to keep small groups working productively.

A.5.4 Recorders

Most of the training meetings described in this book do not specifically call for a person to record the proceedings of the meeting. Occasionally, a meeting description suggests that points could be noted on a chalkboard or overhead transparency. This suggestion is sufficiently infrequent that the meeting leader can probably handle the added responsibilities. The role of recorder is listed here for others who may wish to develop training meetings that would profit from the presence of a recorder.

Preparing for the Meeting

The meeting recorder

- reviews the agenda before the meeting,

- confirms attendance at the meeting or contacts the leader *well ahead of the meeting* to make alternative arrangements, and

- does whatever preparation is required.

Conducting the Meeting

The meeting recorder

- captures ideas without editing or paraphrasing,

- checks to make sure that information has been recorded correctly, and

- helps to keep track of what has been accomplished in the meeting.

A.6 Dealing With Difficult Situations and People in Meetings

Running a meeting often means exerting some kind of control over the behavior of meeting participants via interpersonal means. For example, some meeting participants feel good about themselves when they do almost all of the talking while stifling others. Other participants with good ideas refuse to contribute these ideas to the meeting. The purposes of this section are to point out some of the more common difficulties, and to suggest ways that meeting leaders and facilitators may try to address the situations using interpersonal skills. An excellent reference for trainers who wish to develop interpersonal skills is by Carnegie [22].

A.6.1 Common Problems for Meetings

The problems described here are listed in many references on meetings, and further suggestions for dealing with the situations described here can be readily found. Alternatively, see Adams [1] for a humorous look at how to create the problems.

1. Side conversations

   Problem: Participants are involved in too many side conversations that disrupt the meeting.

   Why?

   - Sometimes participants think that different items should appear on the agenda, so they begin to discuss their point of view with neighbors.

   - Sometimes participants will have a contribution to make that is overlooked, perhaps because of time concerns. Nevertheless, the participants think that their point is important and want to express it to someone.

   - Participants may simply be bored with the meeting, so they start talking to neighbors to pass the time.

   - Occasionally, there will be an individual who needs the constant attention of others, and this person will start conversations to generate this attention.

   Solutions: Many of the strategies that classroom teachers use to deal with students who persistently talk out of turn are applicable here. Several suggestions are offered below.

   - Stop the meeting, and ask the participant, "Do you have an idea that you would like to share with the rest of us?"

   - While the meeting progresses, walk around the room, and take up a position beside the talkers.

   - Ask the participant a question about a recently made point, or ask the participant to comment on the point currently under discussion.

   - Call on the talker(s) by name. Ask them if there is anything that they would like to add to the meeting. If the answer is "No," remind the talkers that side conversations can be very distracting to people who are trying to pay attention to the meeting.

2. Participants who dominate the meeting

   Problem: A participant who wants to do all the talking in the meeting, or who wants to direct what goes on in the meeting.

   Why?

- Some participants seem to need an unusual amount of attention from others. Dominating the meeting may be their way of satisfying that need.

- Instructors who are very enthusiastic about their teaching methods often want to share their ideas with others. Although well intentioned, the enthusiastic instructor may wind up doing all the talking.

- Some people who do not prepare properly for meetings try to cover this by doing a lot of talking. Sometimes it is clear that the participant has not prepared properly.

- Some participants who feel that they have extensive knowledge or experience with a particular topic may talk at length about it.

- Participants who have a great deal of personal authority, perhaps due to their position or rank within the institution, sometimes feel that they have the right to speak a lot at meetings.

Solutions: Many of the strategies used to deal with students who persist with individual questions or concerns, while the rest of the class is ready to move on, will be helpful when dealing with dominating meeting participants.

- When possible, for example if the speaker pauses, thank the speaker for his or her contribution, and use the agenda or the timetable to get the meeting back on track.

- If direct action does not seem appropriate, use body language like glancing at your watch, shuffling or arranging papers, or anything else that suggests that you are either concerned about the time or not paying a lot of attention to the speaker. Be warned that sometimes this kind of signal can make the speaker angry and determined to keep speaking as long as possible.

- Ask other participants for their thoughts. This is probably not going to save any time, but it will at least get others contributing to the meeting.

- Acknowledge the participant's contribution, and point out that meeting time is limited. Ask the participant if he or she would like to continue the conversation after the meeting.

3. Disagreeable participants

   Problem: A participant who thinks he or she knows it all, or who delights in quibbling with the meeting leader.

   Why?

   - Since participants have lives outside the meeting room, there is the chance that a participant is upset by something else that has happened.

   - Naturally relaxed, laid-back meeting participants may be offended or aggravated by some of the specific issues or opinions presented in the meeting.

   - Some people are just show-offs and like to create a spectacle by disagreeing.

   - Some people have combative personalities and feel that they have a right to dispute or challenge everything that is said. While it is not a bad thing for participants to challenge assumptions or suggestions that they think absurd, people with combative personalities take it several steps too far.

   - Some participants who are shy or unsure of themselves sometimes state their opinions in a very confrontational way, to try to compensate for their lack of self-confidence.

   - Participants will sometimes become disagreeable if they feel that they are being ignored.

   Solutions: Many experienced instructors will have dealt with disagreeable students. Most instructors try to be courteous, considerate, and objective with students. This is also appropriate for dealing with disagreeable participants in meetings. Several suggestions are listed below.

   - If the participant has used "loaded" language when making comments, try to paraphrase the participant's comments *without* using loaded or emotional language. The participant will often calm down, agree, and restate his or her position in objective terms.

- Respond to the *content* of the participant's statement. Ignore the attack. Some people feel that they have to respond to attacks with defenses or counterattacks. It is not always the case, but many meeting participants will respect a meeting leader or facilitator who can keep cool during such an episode.
- Find some merit in one of the participant's comments, and then move on.
- Remind the group that meeting time is limited. If people are interested, then the discussion can be continued later.

4. Shy, quiet, or distracted participants

Problem: A participant who is not participating to the extent that the meeting leader or facilitator feels appropriate.

Why?

- Boredom, indifference, or lack of understanding of the topics at hand can result in a lack of participation.
- Busy people may be distracted by other concerns or events in their lives, and they may not participate fully as a result.
- Some people are just not very sure of themselves, and so they are reluctant to participate.
- Some people feel that they already know all that there is to know about the topic at hand, and therefore have no incentive to participate.
- Sometimes, past conflicts can make participants reluctant to talk or work together.

Solutions: Instructors who have used cooperative learning in classes, especially in classes where the students have little faith in their ability or little experience with the subject matter, will have run into similar situations. Many of the techniques used to encourage students to participate in cooperative learning groups will be relevant to encouraging quiet meeting participants. Some suggestions are offered below.

- Call on reticent participants by name, and either ask them a question, or ask them to share their perspectives on the topic under discussion.
- If the problem appears to be a fear of speaking in front of the whole group, have the participants discuss their views with a neighbor before speaking in front of the whole group.
- If there is a break or other semi-private opportunity, take the participant aside and ask if there is any reason why he or she is so reticent.
- If the participants have broken up into groups, and some people are not contributing to their group, ask the non-participators to summarize the group's findings or results.

5. Antagonistic participants

Problem: Participants who are skeptical about the value of the meeting, the use of time, or the relevance of the subject matter.

Why?

- Many participants will have been to meetings that were not effective. They assume, based on that experience, that the training meetings cannot be effective either.
- The leader might not have circulated an agenda or opened the meeting with a statement of what the meeting is intended to achieve.
- The participant might have had some run-ins with the meeting leader.
- The participant may be used to operating with a great deal of personal freedom, and resent the "imposition" of this meeting.
- Very busy people are legitimately concerned about how their time is used. If the participant feels that he or she could accomplish more by not attending the meeting, then antagonism can result.

Solutions: Antagonistic participants are often difficult to deal with directly in the meeting. Sometimes a participant will exhibit antagonistic behavior *because* he or she was dealt with directly in another meeting. As with antagonistic students, it is usually best to meet with antagonistic participants one-on-one.

- Be sure to circulate agendas before the meeting.

- Always start the meeting with a statement of what the meeting will achieve.

- Before calling a meeting, make sure that it's really needed.

- If the antagonism is due to personal factors, get together with the antagonistic participant after the meeting to try to work out the differences.

- Try to find one thing of value in the participant's comments, paraphrase without using loaded language, then move on. By acknowledging the participant, you may be able to calm him or her down.

- Point out that a person who is constantly disputing or criticizing everything that is said is putting the meeting behind plan. Since the meeting time is limited, ask the participant to continue the discussion after the meeting, as everyone else is ready to move on.

6. Prima Donnas

   Problem: Participants who routinely expect special or exceptional treatment.

   Why?

   - Some people have unrealistically high opinions of themselves.

   - Many people have never had to work with anyone that they didn't select themselves, for example they are used to choosing co-authors or research partners. They may have a lack of appreciation for what it takes to work cooperatively with people with whom they would not normally choose to work.

   - Most people have complicated lives and appreciate all of the consideration that they can get. Some people go too far, and believe that they are entitled to all the consideration they want.

   - Some people push the saying, "There's no harm in asking," to the limit. They routinely ask for privileges or considerations that are inappropriate, although they don't always expect to get what they ask.

   Solutions: These individuals can be difficult to deal with. They are often quite easily offended. Usually, arguments based on the statement "We're all in this together—we all need to do our part," are ineffectual, because the prima donna sees himself or herself as different or special.

   - In the short term, it is usually easiest just to cave in to the prima donna's requests, but this can have serious consequences in the long term. For example, other meeting participants may notice the special treatment and demand it for themselves.

   - Make sure that the prima donna has received, and read, the description of the responsibilities of a meeting participant. Ask the prima donna to explain how his or her special request will contribute to their role of meeting participant.

   - During the meeting involve a lot of people, including the prima donna, in discussion.

   - Assign the prima donna a job to complete during the meeting. For example, ask the prima donna to be the recorder.

A.6.2   Confrontations in Meetings

Confrontations in meetings are the result of conflict. In the meeting entitled, "Dealing with Difficult Instructor—Student Situations," suggestions are offered for dealing with conflict. Confrontations in meetings may be dealt with using the process for resolving conflict suggested in "Dealing with Difficult Students."

Confrontations in the training meetings described in this book are usually more difficult to handle than conflicts with students outside of class. When a confrontation occurs during a training meeting, there is an audience. In addition, meeting time is usually very limited. If it is possible, it is usually best to try to acknowledge that the conflict exists, promise to resolve it when the meeting is over, and then move on. After the meeting, be sure to follow up with the participants, to sort out the situation. This has the advantage that the conflict can be worked out in a more private setting, and other participants' time is not wasted. Fortunately, full-blown confrontations in training meetings are not very common.

A.6.3 Why do Difficult Situations Happen at Meetings?

Tropman [114] reports complaints about meetings from meeting *participants*. Some of these comments are given below.

- No one is leading our meeting.

- We float from point to point without any guidance or direction.

- The chair does not chair.

- The chair is a footstool.

Complaints from meeting *leaders* are also given by Tropman [114]. Some of these comments are given below.

- No one says anything.

- When I ask what anyone thinks, half the people look at the ceiling, and the other half look at the floor.

- I thought I had entered Madame Tussaud's wax museum.

The participants' comments clearly indicate that they expect the leader to take charge of the situation in some way, although there is little indication of exactly what they expect the leader to do. On the other hand, the leaders' comments clearly indicate that they expect the participants to *participate* in the meeting, and that this participation will, at least in part, guide the course of the meeting. The participants are either unaware of this expectation, or resistant to participating in this way. Either way, their reaction is passive, and more than likely, they hold their tongues during the meeting and complain bitterly afterwards. Ultimately, the result is that neither the leader's nor the participants' expectations are satisfied, and no one is happy with the meeting. Although it is easy to implicitly assume that all meeting participants will be cognizant of the role that they are expected to play, this is not the case, hence the importance not only of assigning roles for meetings, but of clearly communicating the responsibilities of those roles.

Tropman [114] analyzes many of the hidden functions of a meeting, and the factors that contribute to the failure of many meetings. One factor that is particularly relevant to training meetings in the academic settings is the individualistic focus of most of the people involved in the meeting. Zander [123] makes the following statement.

> Individuals feel that the organization should help them; it is not the individual's prime job to help the organization. [B]asic values ... foster the formation of groups that put the good of the individual before the good of the group.

Based on this statement, Tropman suggests that there is a tendency to single out and blame individuals when things go wrong with meetings. Instead of looking for corporate, collective, or systemic elements in behavior, individual acts are targeted and singled out for intervention.

The primarily individualistic focus of meeting participants can also contribute to the difficulties encountered in meetings in more direct ways. For example, many meeting participants approach meetings with questions like, "What am *I* going to get out of this?" Far fewer meeting participants seem to ask themselves, "What do I have to contribute to this?"

Another direct effect of the individualistic focus of meeting participants is a diminished sense of *active participation*. With the implicit assumption that the group exists to serve the individual, the individual has no responsibility to contribute to the meeting. Successful training meetings usually involve an exchange of ideas among the meeting participants. When many meeting participants feel that all they have to do is sit back and take whatever appeals to them, the potential for the exchange of ideas is substantially diminished.

Lastly, if individuals believe that the group exists to serve them, they may be less inclined to restrain the difficult, often anti-social, behavior described in this section.

A.7   Improving Meetings

The meetings presented in this book represent a component of an ongoing program of professional development. Given the variation in the needs and abilities of meeting participants, it is unusual to immediately find exactly the right way to run a meeting. This section aims to suggest ways that meetings can be refined to cater to the needs and abilities of the participants.

In order to make meetings more effective, it is necessary to gather information from participants. This information is usually most valuable if

1. the information is collected as soon after the meeting as possible,

2. the participants are as specific as possible,

3. the participants supply examples to illustrate their points,

4. the participants respond honestly, and

5. the people running the meeting take the information seriously and act on it.

The last point is very important. If meeting leaders are not motivated to change their meeting practices, then collecting feedback can actually create negative feeling among meeting participants when they see that their suggestions are not being taken seriously.

Haynes [57] suggests a model for improvement. This model has been adapted to produce the schematic below.

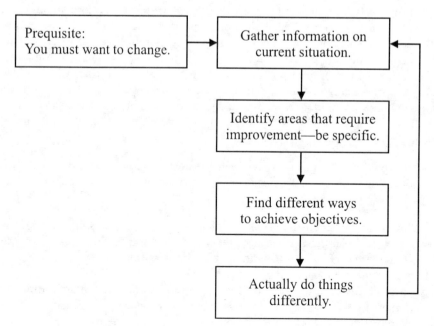

A.7.1   Methods for Meeting Evaluation

Three widely used methods for meeting evaluation are suggested here. All three are readily available for training meetings held in academic settings.

1. Participant Questionnaires

   A sample questionnaire is provided in Section A.9. This can be distributed to participants at the end of a meeting and collected immediately, or the participants can be asked to complete the questionnaire on their own time and return it. The second option sounds ideal, although it should be noted that the rate of return is usually significantly lower. A fifty percent return rate is very high. Furthermore, it is often the participants who were least pleased with the meeting, and who may be able to provide the most useful feedback, who do not usually return completed questionnaires. This can severely skew the results.

2. Observation

   If the professional development program includes a class visit component, then class observers could be invited to attend a training meeting, and observe in much the same way that they would conduct an "observational visit." See the chapter on class visits for further details. This can be a helpful source of information, as the observer is often more objective than either the participants or the leader.

   A second option is to have a "feedback visit" in a training meeting. Although such a visit has the potential to develop very useful information, the fact that meeting time is very limited should be taken into account.

3. Self-Evaluation

   This is something that every meeting leader should do, perhaps in consultation with the meeting facilitator, after each training meeting. The self-evaluation can be as simple as answering three questions.

   (a) What went well?
   (b) What didn't go well?
   (c) What am I going to do about it?

   A self-evaluation is probably the minimum evaluation that a leader who is really interested in improving training sessions should do.

A.8   An Alternative to Large Meetings: Focus Groups

In this context focus groups refer to relatively short meetings with relatively few participants. These meetings include a leader and three to five participants, and they last for about six minutes per participant.

Focus group meetings differ from the training meetings described in this book in that they do not have a "syllabus" to be "covered" during the meeting. Instead, each participant has a few minutes to describe his or her current classroom situation and receive ideas from other participants. Typically, meeting participants are asked to prepare a two to four minute summary, including

- what went well this week in the participant's class, and

- what the participant wasn't satisfied with.

The other meeting participants are then asked to offer ideas for addressing the specific difficulties that have been raised.

Focus group meetings can be a lot of fun, and a lot of valuable information can be generated. Some instructors prefer focus groups to more traditional training sessions, because of the individual attention and the more relaxed atmosphere. However, it is important to point out that focus group meetings are not a highly effective way to communicate information, and that the quantity of ideas generated can be limited by the small number of participants.

The effects of difficult participant behavior are different for focus groups. If a difficult person represents 25–30% of the meeting participants, then the disruptive effects of this person can be even more pronounced. On the other hand, fewer people are affected by the disruptive behavior. A great deal of disruptive behavior stems from the conflict between individual identity and membership in a group of meeting participants, as was noted in Section A.6.3. With fewer participants and an opportunity for each individual to speak, there is usually less need for participants to assert their individuality through disruptive behavior in a focus group setting.

A.9 Meeting Evaluation Questionnaire

Meeting: _____

For items 1 to 7, please circle the number that most accurately reflects your opinion. For items 8 to 11, details and examples will help to illustrate your opinions.

		SD 1	D 2	N 3	A 4	SA 5
1.	The meeting achieved its objectives.	1	2	3	4	5
2.	My personal objectives for this meeting were achieved.	1	2	3	4	5
3.	The items on the agenda were relevant.	1	2	3	4	5
4.	People stayed focused on the items on the agenda.	1	2	3	4	5
5.	The meeting was well prepared.	1	2	3	4	5
6.	Meeting time was used well.	1	2	3	4	5
7.	The length of the meeting was appropriate.	1	2	3	4	5

SD Strongly disagree   **D** Disagree   **N** Neutral   **A** Agree   **SA** Strongly Agree

8. Which portions of the meeting were most useful to you?

- _____

- _____

- _____

- _____

- _____

9. Which aspects of the meeting were least useful to you?

- _____

- _____

- _____

- _____

10. What actions will you be taking as a result of this meeting?

- _____

- _____

- _____

- _____

- _____

11. Please list other comments or suggestions.

- _____

- _____

- _____

- _____

- _____

Appendix B

The Michigan Introductory Program

Both authors were graduate students in the mathematics department at the University of Michigan. They began as graduate student instructors, and were trained in the professional development component of the Michigan Calculus Project. They both taught Precalculus, Calculus I, and Calculus II. Near the ends of their graduate careers, both authors then became instructor trainers. This appendix gives an overview of the Michigan Calculus/Precalculus program, which frames the experiences and backgrounds of the authors.

B.1 Course Goals

The following are the goals of the introductory courses as stated in the "Michigan Introductory Program Instructor's Guide" [104].

Establish constructive student attitudes about math:

1. interest in math
2. value of math, and its link to the real world
3. the likelihood of success and satisfaction
4. the effective ways to learn math

Strengthen students' general academic skills:

1. critical thinking
2. writing
3. giving clear verbal explanations
4. working collaboratively
5. assuming responsibility
6. understanding and using technology

Improve students' quantitative reasoning skills:

1. translating a word problem into a math statement, and back again
2. forming reasonable descriptions and judgments based on quantitative information

Develop a wide base of mathematical knowledge:

1. understanding of concepts
2. basic skills
3. mathematical sense (quantitative, geometric, symbolic)
4. the thinking process (problem-solving, predicting, generalizing)

Improve instruction:

1. more instructor commitment and interest
2. increased student—faculty contact
3. student-centered approach
4. satisfaction with teaching
5. ability to organize and maintain multiple methods of instruction

Improve student persistence rates:

1. students continue taking math and science
2. more students become math majors

B.2   Teaching Methods

Towards achieving the aforementioned goals, the instructors in the introductory courses are encouraged to use a variety of teaching methods, beyond the traditional lecture method. These include

- using interactive lectures and a variety of questioning techniques to involve students in the lectures,
- using cooperative learning activities in class,
- assigning weekly homework to four-person homework teams,
- encouraging and expecting the students to read the textbook,
- requiring mathematical writing,
- integrating hand-held graphing calculator technology into the lessons, and
- emphasizing underlying concepts and real-world problems.

B.3   Typical Classes

An instructor in one of the introductory courses teaches one section of either Precalculus, Calculus I, or Calculus II. These courses are multi-section courses, with twenty-five to sixty sections of a course running concurrently. The classes consist of up to thirty-two students, who are mostly first-year students. The overwhelming majority of the students are not math majors, but are taking calculus as a service course to their major. The classes meet three times per week for ninety minutes per session. The classrooms typically consist of eight to ten four-person tables, with blackboards at either end of the room.

B.4   Sample Lesson Plans

Instructors in their first term teaching in the introductory courses at the University of Michigan are given lesson plans for the bulk of the term. The use of these lesson plans is optional, but strongly encouraged. Roughly two-thirds of the way through the term, the instructors are given a workshop on developing their own lesson plans. They are then required to write lesson plans for roughly a week of classes. After that, lesson plans are again given to the instructors for the remainder of the term.

The following pages contain three sample lesson plans from the precalculus course at the University of Michigan. The textbook used is [31]. The references to problems in the lesson plans are references to problems from [31].

Day: 1

Last night's reading: Course pack

Last night's homework: none

Section: 1.1

Assigned reading: 1.1, 1.2

Assigned homework: 1.1

Goals

- Communicate essential course information

- Instructor and students begin learning each others' names

- Gather information from students (data sheet)

- Set an atmosphere of active learning

- Students learn the definition of function

- Students able to differentiate between functions and non-functions and learn the vertical line test

- Students introduced to reading and interpreting graphs

Before class Write the following on the board. Make sure that your writing is large enough for everyone to see clearly.

- Your name
- Math 105
- Your section number
- The reading and homework assignment, and that you expect it to be completed by the next class

00:10-00:40 Go through first day activities, including:

- Introduce yourself (What should students call you?)
- Distribute handouts and cover course policies and expectations
- Remind students of their responsibilities (attendance, reading, etc.)
- Collect student data sheets
- Explain that they will be doing homework in groups, that the first assignment is due one week from Friday, and that you will assign the groups next class period
- Allow students to ask you questions.

Since working together will be an important part of the course, the students should get to know each other as quickly as possible. Therefore, if you feel comfortable with it, you should include an opportunity for students to get to know each other. (For example, you could get the class in a circle and play the "name game." Make sure you go last. Another idea is to ask them to pair up with someone they don't know, and find out two words which describe how that person feels about taking Math 105. When you are ready, call their attention to you and have each student introduce their partner and say the two words aloud.)

00:40-00:45 Ask the class if anyone knows what the mathematical definition of a function is. Facilitate a discussion that culminates with an acceptable definition written on the board.

00:45-00:50 Tell the class that one of the features of this course will be a lot of active work in class, so that they can get hands-on experience with concepts and calculations. Have the students pair up with someone close to them. In their pairs have them answer 1.1 #1 page 6. After a couple of minutes, call on some individuals to explain their answers. Make sure "each input has a unique output" makes sense to them.

00:50-00:70 Do the group activity "Section 1.1 The Graph of a Function and the Vertical Line Test," which is attached. Break the class into teams by counting off, so that they get used to moving around the room.

00:70-00:80 In the same groups as before, have the students work through 1.1 #7 on pages 8-9. Assign each group one of (b)(i), (b)(ii), or (c). Ask them to choose a scribe, who will draw their graphs and write their reasons on the board.

00:80-00:88 When there is at least one answer for each of (b)(i), (b)(ii), and (c) on the board, call the attention of the entire class to the board. Have someone other than the scribe explain their answers at the board.

00:88-00:90 Remind class of the reading and homework assignment. Stress that it is critical for them to read the book before they come to class, because you will not be covering everything in the reading.

Day: 8
Last night's reading: 2.3
Last night's homework: 2.2

Section: 2.3
Assigned reading: 2.4
Assigned homework: 2.3

Goals

- Students understand geometric effects of parameters in $y = mx + b$

- Students able to find intersections of two lines, and know why it may be important to do so

- Students able to recognize horizontal and vertical lines, and when two lines are parallel or perpendicular

Before class Write the following on the board:

- Bring calculators and cords to class next time

- The reading and homework assignment for next hour.

- Collect the team homework.

- Return the graded quiz.

00:10-00:15 Hand back the quiz, and quickly sketch the answers. You may want to pay closer attention to anything that most of the class missed. Tell them to see you in office hours if they have questions about the quiz. They will want you to give them an idea of the distribution. Tell them the mean or median, but don't reveal the high and low scores. Remember not to debate grading in class.

00:15-00:25 Ask your students to take three minutes to discuss with a partner any difficulties they might have had with the reading or homework problems from last night. Ask them if there are issues that they were not able to clear up with their partner. If there are any questions, address two or three. Encourage them to bring unresolved questions to you in office hours or to the Math Lab.

00:25-00:35 Interactive lecture on how the parameters in the slope-intercept equation affect the appearance of the graph. You could make a table:

	Slope (m) positive	Slope (m) negative
Intercept (b) positive		
Intercept (b) negative		

The entries in the table would consist of an example with appropriate slope and intercept, and the graph of that equation. You should also discuss how lines with m less than 1 and m greater than 1 differ in their graphs.

00:35-0:50 In groups, have them work out 2.3 #20 page 96 parts (a) and (b). In part (b), make sure they try to reason which graph is which based on the parameters calculated in part (a). You can wrap this up by getting one person to write their equations on the board, while another draws the graphs. Have the graph drawer explain which equation goes with which graph, and why.

00:50-00:55 Draw their attention to part (c) of the previous exercise. Without working it out, discuss how to tell which company is cheapest for a certain number of minutes. Use the conversation to highlight the importance of finding intersections of lines in application problems.

00:55-00:65 In their groups, conduct a roundtable exercise on finding intersections of lines. Have the groups take out a piece of paper. Have them pass the paper around the group, while the students in the group each write down the equation for a line. Remind them that there is more than one general form, and encourage them to use all they know. Now pass the papers back around the groups. Each student in turn is to choose two of the lines on the paper, and find their intersection. The other students in the groups should monitor the progress of the intersection-finders. Monitor group progress throughout the room, and remind them that they can accomplish this task algebraically or by tracing on their calculators.

00:65-00:75 Give a mini-lecture on parallel lines, perpendicular lines, horizontal lines, and vertical lines. Be careful when dealing with the algebraic condition for perpendicular lines. A horizontal line is clearly perpendicular to a vertical line, but the algebra doesn't work.

00:75-00:80 Preview the next section on regression lines. Ask them to suppose that they are working for a consulting firm analyzing sales data for a company, or a government bureau analyzing some other real-world phenomenon. Discuss why data gathered could have a linear trend, but why it would be extremely unlikely that the data points would lie exactly on a line. Another idea is to relate an experiment you have done. You could outline the experiment, and show a scatter-plot of the data you collected. Tell them that the next section deals with finding the best line to fit sets of data that have a linear trend, but do not reveal an exact linear relationship.

00:80-00:90 Catch-up time if you are behind, or use the time to take questions. Remind class of the reading and homework assignment, and that they need to bring their calculators and cords next time. Tell your students to have a good weekend, and thank them for another week of hard work.

Day: 22
Last night's reading: 4.4
Last night's homework: 4.3

Section: 4.4
Assigned reading: 4.5
Assigned homework: 4.4

Goals

- Students know the general formula for quadratic functions in both vertex form and standard form

- Students able to convert between vertex and standard form, including the ability to complete the square

- Students able to convert between graph and formula for quadratic functions

Before class Write the following on the board:

- The reading and homework assignment for next class

00:10-00:20 Go over individual homework as usual.

00:20-00:45 Quiz, if this is your day to do it.

00:45-00:50 Mini-lecture on the vertex form of a quadratic function (see the summary in the box page 233). Point out how any quadratic function is obtainable from $y = x^2$ by a horizontal shift, followed by a vertical stretch (with maybe a flip), followed by a vertical shift. Use the three examples they worked on from last class to illustrate the general principles. Point out how easy it is to obtain the vertex and axis of symmetry from the vertex form.

00:50-00:60 Divide the class into groups of three, and have them work out 4.4 #4 page 239. When you wrap up, be sure to emphasize how to incorporate the information given in the problem into the graph and formula. When going over part (d), emphasize how the symmetry of the parabola lets you find where the second zero is. Those that finish early can work on 4.4 #5 page 239.

00:60-00:65 Mini-lecture on converting from vertex form to standard form by multiplying out (they will call it FOIL–first outside inside last), and on converting from standard form to vertex form by completing the square. You can again refer to the examples done last class.

00:65-00:75 Have the students pair up. Have one student in each pair do 4.4 #29 page 242 while their partner monitors and assists if needed. Then have the other student in each pair do 4.4 #30 page 242 in the same manner.

00:75-00:88 In groups have them do 4.4 #41 pages 243—244. Require them to put the function into vertex form by completing the square, and then graph the function without the aid of the calculator. Make sure they work out part (b) algebraically, rather than by graphing and tracing on their calculators. In parts (c) and (d), get them to find the height and horizontal distance traveled by realizing that the highest point is the vertex. They should only use their calculators to check themselves. Those that finish early can work on 4.4 #14 page 241.

00:88-00:90 Remind class of the reading and homework assignment, and that the team homework is due next time.

B.5   Instructor Responsibilities

An instructor for an introductory course at the University of Michigan is one of many instructors in the uniform system. Each instructor is responsible for his or her own class, but some of the typical teaching duties are held by a team of one to three course coordinators. The breakdown of the responsibilities between the instructor and the course coordinator is from the "Michigan Introductory Program Instructor's Guide" [104].

The instructor's responsibilities include:

- Explaining the course's goals to the students
- Planning and carrying out the classroom instruction using a variety of teaching methods including lecturing, active learning activities, going over homework, etc.
- Assigning and grading homework
- Writing and grading quizzes
- Participating in proctoring and grading uniform midterm and final exams
- Assigning final course grades (according to uniform guidelines)
- Holding office hours (3 hrs. weekly minimum), with one of the hours held in the Math Lab
- Attending weekly course staff meetings
- Conducting all classes personally except in an emergency; notifying the coordinator in an emergency
- Notifying the course coordinator of any problems related to teaching
- Delivering all course announcements to students
- Beginning promptly
- Responding to E-mail quickly

The responsibilities of the course coordinator include:

- Providing a course syllabus and assignments
- Giving guidelines for the pace of the course
- Planning and running the course staff meetings
- Providing assistance with teaching
- Visiting and observing classes
- Conducting midterm small-group feedback sessions with students
- Writing uniform midterms and final exams
- Directing the uniform exam procedures
- Direct uniform exam grading sessions
- Assigning letter grades for uniform exams
- Handling student complaints
- Approving the instructors' final course grades

The course coordinators play a role that is not only administrative, but if they are coordinating a course with first-time instructors, they lead in the professional development of the new instructors. For this reason, they are also sometimes referred to as "mentors" or "trainers." When this is the case, the course coordinators have the additional unlisted duties of overseeing the professional growth of new instructors, encouraging instructor interest in teaching, helping instructors develop a balance between teaching and research duties, and maintaining an open door for consultation on teaching issues.

B.6 Typical Professional Development Program

The instructors who are new to teaching in the introductory program at The University of Michigan participate in a professional development program similar to the one described in Chapter 1. During the week immediately prior to the beginning of classes, they participate in the professional development week, which is explained in [103]. During this week they are introduced to the course philosophy and goals, given information on administrative support mechanisms, and immersed in some of the immediate activities of teaching a course in the manner required. For example, there is practice lecturing, practice designing and leading a cooperative learning activity, and practice asking and answering questions. During the semester the instructors attend weekly course meetings, which serve the dual functions of dispensing course information and expanding on the professional development started during the training week. During the semester, the course coordinators conduct one observational visit and one student feedback session for each new instructor. If there are difficulties with a particular instructor, more observational visits are done. After the one semester of intensive assistance, the instructors continue their careers at the university by teaching a variety of courses in the introductory program, or running computer lab sections for one of the intermediate courses. In addition to the professional development program required by the department, other optional opportunities for pedagogical growth include the new Educational Issues Seminar in the Mathematics Department and the many discipline-independent teaching workshops conducted by the Center for Research on Learning and Teaching.

Bibliography

[1] S. Adams. *The Dilbert Principle*. HarperCollins Publishers, Inc., New York, 1996.

[2] H. L. Amick. Math class–have you seen the preview? In the *Innovative Teaching Exchange*, 1998. http://706.4.57.253/t_and_l/exchange/ite1/ite1.html

[3] J. Andersen. *Teaching Students to Read Technical Material: The Use of Reading Outlines*. Harcourt Brace, Orlando, 1996.

[4] T. Angelo and P. Cross. *Classroom Assessment Techniques*. Jossey–Bass, San Francisco, 2nd edition, 1993.

[5] F. Avenoso *et. al. Instructor's Manual with Sample Exams to accompany Calculus, by Hughes-Hallet, Gleason, et. al.* John Wiley & Sons, New York, 1994.

[6] S. Biaocco and J. DeWaters. *Successful College Teaching. Problem Solving Strategies of Distinguished Professors*. Allyn and Bacon, Boston, 1998.

[7] M. C. Barrett and S. Montgomery. Undergraduate women in science and engineering: Providing academic support. *CRLT Occasional Papers*, 8:1–4, 1997.

[8] W. Berquist and S. Phillips. *A Handbook for Faculty Development. Volume I.*, The Council for the Advancement of Small Colleges, Danville, NY, 1975.

[9] B. Black, M. Brown and P. Shure. *Michigan Introductory Program Instructor's Guide*. The University of Michigan Department of Mathematics, Ann Arbor, MI, 1997.

[10] L. Blake and M. Reed, editors. *Duke University Laboratory Calculus Manual*. John Wiley and Sons, New York, 1999.

[11] K. Blanchard and S. Johnson. *The One Minute Manager*. Berkley Books, New York, 1981.

[12] B. S. Bloom, editor. *Taxonomy of Educational Objectives. Handbook I: Cognitive Domain*. Longmans Green, New York, 1956.

[13] W. Blume and M. K. Heid. *Teaching and Learning Mathematics with Technology*. Pennsylvania State University, University Park, PA, 1997.

[14] M. Boelkins and T. Ratcliff. How we get our students to read the text before class. *FOCUS*, 21(1): 16–17, 2001.

[15] J. Bookman and L. Blake. Seven years of Project CALC at Duke University: Approaching a steady state? *PRIMUS*, 6(3): 221–234, 1996.

[16] J. Bookman and C. Friedman. A comparison of the problem solving performance of students in lab based and traditional calculus. *CBMS Issues in Mathematics Education*, 5: 101–116, 1994.

[17] R. Borelli and C. Coleman. *Differential Equations: A Modeling Perspective*. John Wiley and Sons, New York, 1998.

[18] W. Boyce and J. Ecker. The computer-oriented calculus course at Rensselaer Polytechnic Institute. *The College Mathematics Journal*, 26(1): 45–50, 1995.

[19] C. Burnap and M. Leiva. Fostering collective responsibility for student learning: Teaching seminars at the University of North Carolina at Charlotte mathematics department. in P. Hutchings, editor. *Making Teaching Community Property: A Menu for Peer Collaboration and Peer Review.* American Association for Higher Education, Washington, D.C., 1996.

[20] W. S. Bush. *Preservice Secondary Mathematics Teachers' Knowledge about Teaching Mathematics and Decision Making During Teacher Training. (Doctoral Dissertation, University of Georgia, 1982.)* Dissertation Abstracts International, 43, 2264A, 1983.

[21] J. Cangelosi. *Classroom Management Strategies.* Longman, White Plains, NY, 3rd edition, 1997.

[22] D. Carnegie. *How to Win Friends and Influence People.* Simon and Schuster, Inc., New York, 1982.

[23] B. Case, editor. *Responses to the Challenge. Keys to Improved Instruction from Teaching Assistants and Part-Time Instructors. MAA Notes #11.* Mathematical Association of America, Washington, D.C., 1985.

[24] B. Case, editor. *You're the Professor, What Next? Ideas and Resources for Preparing College Teachers. MAA Notes #35.* Mathematical Association of America, Washington, D.C., 1994.

[25] R. Cava. *Difficult People.* Firefly Books, Inc., Buffalo, 1990.

[26] Indiana University Audio Visual Center. *Observing Teaching.* From the series: Strategies on College Teaching. 1994.

[27] R. Chang. *Success Through Teamwork.* Richard Chang and Associates, Inc. Publications Division, Irvine, CA, 1994.

[28] R. Chang and K. Kehoe. *Meetings That Work!* Richard Chang and Associates, Inc. Publications Division, Irvine, CA, 1996.

[29] M. Chesler. Perceptions of faculty behavior by students of color. *CRLT Occasional Papers*, 7: 1–7, 1997.

[30] S. J. Coffman. Improving your teaching through small–group diagnosis. *College Teaching*, 39(2): 80–82, 1991.

[31] E. Conally *et. al. Precalculus: Functions Modeling Change (Preliminary Edition).* John Wiley & Sons, New York, 1998.

[32] L. Cooley. Evaluating student understanding in a calculus course enhanced by a computer algebra system. *PRIMUS*, 7(4): 308–316, 1997.

[33] C. Cowen. Teaching and testing mathematics reading. *The American Mathematical Monthly*, 98(1): 50–53, 1991.

[34] A. Crannell. How to grade 300 mathematical essays and live to tell the tale. *PRIMUS*, 4(3): 193–204, 1994.

[35] N. Davidson. Small group learning and teaching in mathematics. In E. Dubinsky, D. Matthews and B. Reynolds, editors. *Readings in Cooperative Learning for Undergraduate Mathematics. MAA Notes #44.* Mathematical Association of America, Washington, D.C., 1997.

[36] B. G. Davis. *Tools for Teaching.* Jossey–Bass, San Francisco, 1993.

[37] M. DeLong and D. Winter. Novice instructors and lesson planning. *Preprint.*

[38] M. DeLong and D. Winter. Addressing difficulties with student-centered instruction. *PRIMUS*, 8(4): 340–364, 1998.

[39] M. DeLong and D. Winter. An objective approach to student-centered instruction. *PRIMUS*, 11(1): 27–52, 2001.

[40] J. C. Derderian and E. Rodriquez-Carrington. Undersampled sine waves. *The College Mathematics Journal*, 29(3): 213–218, 1998.

[41] A. Donellon. *Team Talk: The Power of Language in Team Dynamics*. Harvard Business School Press, Cambridge, MA, 1996.

[42] R. Douglas, editor. *Towards a Lean and Lively Calculus*. *MAA Notes #6*. Mathematical Association of America, Washington, D.C., 1986.

[43] E. Dubinsky, D. Matthews and B. Reynolds, editors. *Readings in Cooperative Learning for Undergraduate Mathematics*. *MAA Notes#44*. Mathematical Association of America, Washington, D.C., 1997.

[44] U. Dudley, editor. *Resources for Calculus Collection. Volume 5: Readings for Calculus*. *MAA Notes #31*. Mathematical Association of America, Washington, D.C., 1993.

[45] P. Dunham and T. Dick. Research on graphing calculators. *Mathematics Teacher*, 87(6): 440–445, 1994.

[46] M. A. Farrell and W. A. Farmer. *Secondary Mathematics Instruction: An Integrated Approach*. Janson, Providence, 1988.

[47] R. Felder. Who needs these headaches? Reflections on teaching first-year engineering students. *Success 101*, Fall 1997.

[48] R. Felder and R. Brent. Cooperative learning in technical courses: Procedures, pitfalls and payoffs. *ERIC Document Reproduction Service Report ED377038*, 1994.

[49] R. Felder and R. Brent. Navigating the bumpy road to student-centered instruction. *College Teaching*, 44(2): 43–47, 1996.

[50] J. Fenty. Knowing your students better: A key to involving first-year students. *CRLT Occasional Papers*, 9: 1–7, 1997.

[51] D. Finkel and G. S. Monk. Teachers and learning groups: Dissolution of the Atlas complex. in C. Boulton and R. Y. Garth, editors. *Learning in Groups*. Jossey–Bass, San Francisco, 1983.

[52] C. Frank. *Ethnographic Eyes: A Teacher's Guide to Classroom Observation*. Heinemann, Portsmouth, NH, 1999.

[53] J. Garofalo. Beliefs and their influence on mathematical performance. *Mathematics Teacher*, 82(7): 502–505, 1989.

[54] B. Gold. Requiring student questions on the text. In the *Innovative Teaching Exchange*, 1998. http://706.4.57.253/t_and_l/exchange/ite3/reading_gold.html

[55] B. Gold, S. Z. Keith and W. A. Marion, editors. *Assessment Practices in Undergraduate Mathematics*. *MAA Notes #49*. Mathematical Association of America, Washington, DC, 1999.

[56] G. Gopen and D. Smith. What's an assignment like you doing in a course like this? Writing to learn mathematics. *The College Mathematics Journal*, 21(1): 2–29, 1990.

[57] M. Haynes. *Effective Meeting Skills. A Practical Guide for More Productive Meetings*. Crisp Publications, Inc., Menlo Park, CA, 1997.

[58] M. K. Heid. Calculators on tests—one giant leap for mathematics education. *Mathematics Teacher*, 81(9): 710–713, 1988.

[59] M. K. Heid. Resequencing skills and concepts in applied calculus using the computer as a tool. *Journal for Research in Mathematics Education*, 19(1): 3–25, 1988.

[60] N. Hewitt and E. Seymour. *Talking About Leaving: Why Undergraduates Leave the Sciences.* Westview Press, Boulder, CO, 1996.

[61] D. Hughes-Hallet, A. Gleason, *et. al. Calculus.* John Wiley & Sons, New York, 2nd Edition, 1998.

[62] C. Jackson and J. Leffingwell. The role of instructors in creating math anxiety in students from kindergarten through college. *Mathematics Teacher*, 92(7): 583–586, 1999.

[63] M. Johnson. *Reading Mathematics.* Harcourt Brace, Orlando, 1996.

[64] J. Jones, *et. al.* Offer them a carrot: Linking assessment and motivation in developmental mathematics. *Research and Teaching in Developmental Education*, 13(1): 85–91, 1996.

[65] P. Judson. Elementary business calculus with computer algebra. *Journal of Mathematical Behavior*, 9(2): 153–157, 1990.

[66] P. Judson. A computer algebra laboratory for calculus 1. *Journal of Computers in Mathematics and Science Teaching*, 10(4): 35–40, 1991.

[67] Z. Karian, editor. *Symbolic Computation in Undergraduate Mathematics Education. MAA Notes #24.* Mathematical Association of America, Washington, DC, 1992.

[68] C. Knapper and S. Piccinin, editors. *Using Consultants to Improve Teaching. New Directions in Teaching and Learning #79.* Jossey–Bass, San Francisco, 1999.

[69] S. Krantz. *How to Teach Mathematics.* American Mathematical Society, Providence, 2nd edition, 1999.

[70] L. Lambert and S. L. Tice, editors. *Preparing Graduate Students to Teach.* American Association for Higher Education, Washington, DC, 1993.

[71] L. Lambert, S. L. Tice and P. Featherstone. *University Teaching. A Guide for Graduate Students.* Syracuse University Press, Syracuse, NY, 1996.

[72] M. Lampert and D. L. Ball. *Teaching, Multimedia and Mathematics. Investigations of Real Practice.* Teachers College Press, New York, 1998.

[73] R. Larson, R. Hostetler and B. Edwards. *Calculus with Analytic Geometry.* Houghton Mifflin, Boston, 6th edition, 1998.

[74] L. C. Leinbach, J. Hundhausen, A. Osterbee, L. Senechal and D. Small, editors. *The Laboratory Approach to Teaching Calculus. MAA Notes #20.* Mathematical Association of America, Washington, DC, 1991.

[75] A. Levine and J. Cureton. College life: An obituary. *Change*, 50(10): 14–17, 51, 1998.

[76] J. Long. Love them into learning. *Teachers in Focus*, 1: 4–8, 1997.

[77] J. Lowman. Promoting motivation and learning. *College Teaching*, 38(4): 136–139, 1990.

[78] P. Maida. Reading and note-taking prior to instruction. *Mathematics Teacher*, 88(6): 470–473, 1995.

[79] R. Mayes. The application of a computer algebra system as a tool in college algebra. *School Science & Mathematics*, 95(2): 69–69, 1995.

[80] W. J. McKeachie. Motivation in the college classroom. *Innovation Abstracts*, 4(12), 1982.

[81] W. J. McKeachie. *Teaching Tips: Strategies, Research and Theory for College and University Teachers.* D.C. Heath and Co., Lexington, MA, 9th edition, 1994.

[82] J. Meier and T. Rishel. *Writing in the Teaching and Learning of Mathematics. MAA Notes #48.* Mathematical Association of America, Washington, DC, 1998.

[83] C. L. Mitsakos. Mirrors for the classroom: A guide to observation techniques for teachers and supervisors. *ERIC Document Reproduction Service Report ED 308620*, 1986.

[84] University of Washington. Small group instructional diagnosis facilitator training. 1981.

[85] Committee on the Teaching of Undergraduate Mathematics. *College Mathematics: Suggestions on How to Teach it*. Mathematical Association of America, Washington, DC, 1979.

[86] A. Osterbee and P. Zorn. *Calculus from the Graphical, Numerical, and Symbolic Points of View*. Harcourt Brace College Publishers, Fort Worth, 1997.

[87] J. Palmiter. Effects of computer algebra systems on concept and skill acquisition in calculus. *Journal for Research in Mathematics Education*, 22(2): 151–156, 1991.

[88] W. G. Perry. *Forms of Intellectual and Ethical Development in the College Years: A Scheme*. Holt, Rinehart and Winston, New York, 1970.

[89] G. Pólya. *How to Solve it*. Doubleday Books, Garden City, NY, 2nd edition, 1957.

[90] T. Ratliff. How I (finally) got my students to read the text. In the *Innovative Teaching Exchange*, 1998.
http://706.4.57.253/t_and_l/exchange/ite3/reading_rattliff.html

[91] M. V. Redmond. A process of midterm evaluation incorporating small group discussion of a course and its effects on student motivation. *ERIC Document Reproduction Service Report ED 217953*, 1982.

[92] M. V. Redmond and D. J. Clark. Small group instructional diagnosis: Final report. *ERIC Document Reproduction Service Report ED 217954*, 1982.

[93] M. Reed and L. Blake. *Instructor's Manual for Math 31L and Math 32L*. Duke University Mathematics Department, Durham, NC, 1999.

[94] A. Reiter. Helping undergraduates learn to read mathematics. In the *Innovative Teaching Exchange*, 1998.
http://706.4.57.253/t_and_l/exchange/ite3/reading_reiter.html

[95] B. Reynolds, N. Hagelgans, K. Schwingendorf, D. Vidakovich, E. Dubinsky, M. Shahin and G. J. Wimbish. *A Practical Guide to Cooperative Learning in Collegiate Mathematics. MAA Notes # 37*. Mathematical Association of America, Washington, DC, 1995.

[96] T. Ricks. *Making the Corps*. Scribner, New York, 1997.

[97] C. Rinne. *Excellent Classroom Management*. Wadsworth Publishing Co., Belmont, CA, 1997.

[98] T. Rishel. *Teaching First. A Guide for New Mathematicians. MAA Notes #54*. Mathematical Association of America, Washington, DC, 2000.

[99] E. Rogers, B. Reynolds, N. Davidson and A. Thomas, editors. *Cooperative Learning in Undergraduate Mathematics. Issues that Matter and Strategies that Work. MAA Notes #55*. Mathematical Association of America, Washington, DC, 2001.

[100] E. Rykken and J. Sorenson. Volumes and history: A calculus project involving reading an original source, 1998. In the *Innovative Teaching Exchange*.
http://706.4.57.253/t_and_l/exchange/ite5/hodgson.html

[101] A. Schoenfeld. When good teaching leads to bad results. The disasters of 'well-taught' mathematics courses. *Educational Psychologist*, 23(2): 145–166, 1988.

[102] A Schwartz. Axing math anxiety. *Education Digest*, 65(5): 62–64, 2000.

[103] P. Shure, B. Black and D. Shaw. *The Michigan Calculus Program Instructor Training Materials*. John Wiley & Sons, New York, 1997.

[104] P. Shure, M. Brown and B. Black. *Michigan Introductory Program Instructor's Guide.* The University of Michigan, Ann Arbor, MI, 1997.

[105] D. Smith. Trends in calculus reform. in A. Solow, editor. *Preparing for a New Calculus. MAA Notes #36.* Mathematical Association of America, Washington, DC, 1994.

[106] M. Solomon. *Working with Difficult People.* Prentice Hall, Upper Saddle River, NJ, 1990.

[107] M. D. Sorcinelli. *Evaluation of Teaching Handbook.* Dean of Faculties Office, Indiana University, Bloomington, IN, 1986.

[108] A. Statham, L. Richardson and J. Cook. *Gender and University Teaching: A Negotiated Difference.* State University of New York Press, Albany, NY, 1991.

[109] A. Sterrett, editor. *Using Writing to Teach Mathematics. MAA Notes #16.* Mathematical Association of America, Washington, DC, 1980.

[110] S. Sullivan and J. Glanz. *Supervision that Improves Teaching: Strategies and Techniques.* Corwin Press, Inc., Thousand Oaks, CA, 2000.

[111] A. Thompson. Teachers beliefs and conceptions: A synthesis of the research. in D. Grouws, editor. *Handbook of Research on Mathematics Teaching and Learning.* Macmillan, New York, 1992.

[112] P. Timm and J. Stead. *Communication Skills for Business and Professions.* Prentice Hall, Upper Saddle River, NJ, 1996.

[113] S. Tobias. Math mental health: Going beyond math anxiety. *College Teaching,* 39(3): 91–93, 1991.

[114] J. Tropman. *Making Meetings Work: Achieving High Quality Group Decisions.* Sage Publications, Inc., Thousand Oaks, CA, 1996.

[115] A. Tucker and J. Leitzel, editors. *Assessing Calculus Reform Efforts: A Report to the Community. MAA Reports #6.* Mathematical Association of America, Washington, DC, 1994.

[116] F. Urbach. Developing a teaching portfolio. *College Teaching,* 40(2): 71–73, 1992.

[117] F. Ward and D. Wilberscheid. *Insight Into Calculus Using the Texas Instruments Graphics Calculators.* Prentice Hall, Upper Saddle River, NJ, 1997.

[118] M. Weimer. *Improving College Teaching.* Jossey–Bass, San Francisco, 1990.

[119] M. Weimer, J. L. Parrett and M. Kerns. *How am I Teaching? Forms and Activities for Acquiring Instructional Input.* Magna Publishing, Inc., Madison, WI, 1988.

[120] A. Whimbey and J. Lockhead, editors. *Problem Solving and Comprehension.* The Franklin Institute Press, Philadelphia, PA, 1980.

[121] D. Winter and C. Yackel. *Graduate Student Instructors and Classroom Authority.* Department of Mathematics, University of Michigan, Ann Arbor, MI, 1998.

[122] D. Winter and C. Yackel. Novice instructors and student–centered instruction: Understanding perceptions and responses to challenges of classroom authority. *PRIMUS,* 10(4): 289–318, 2000.

[123] A. Zander. *Making Groups Effective.* Jossey–Bass, San Francisco, 1982.

[124] S. Zucker. Teaching at the university level. *Notices of the American Mathematical Society,* 43(8): 863–865, 1996.